Strategies of Expertise
in Technical Controversies

Strategies of Expertise
in Technical Controversies
A Study of Wood Energy Development

Frederick Frankena

Bethlehem: Lehigh University Press
London and Toronto: Associated University Presses

Associated University Presses
440 Forsgate Drive
Cranbury, NJ 08512

Associated University Presses
25 Sicilian Avenue
London WC1A 2QH, England

Associated University Presses
P.O. Box 39, Clarkson Pstl. Stn.
Mississauga, Ontario,
L5J 3X9 Canada

The paper used in this publication meets the requirements
of the American National Standard for Permanence of Paper
for Printed Library Materials Z39.48-1984.

Library of Congress Cataloging-in-Publication Data

Frankena, Frederick.
    Strategies of expertise in technical controversies: a study of wood energy development/Frederick Frankena.
        p.   cm.
    Includes bibliographical references and index.
    ISBN 0-934223-14-9 (alk. paper)
    1. Energy power plants—United States—Location.   2. Fuelwood-United States.   3. Energy power plants—United States—Environmental aspects.
4. Expertise.   I. Title.
TK12223F73   1992   89-64068
333.95′39—dc20   CIP

PRINTED IN THE UNITED STATES OF AMERICA

To my wife, Joann

Contents

Figures and Tables

Figures

Tables

Abbreviations

BED   Burlington Electric Department
CRUF   Committee for the Rational Use of Our Forests
DNR   Michigan Department of Natural Resources
EIS   Environmental impact statement
GRP   Grand Rapids Press
LSJ   Lansing State Journal
MW   Megawatt—a measure of power equivalent to 1 million watts or 1 thousand kilowatts
NIMBY   Not-in-my-backyard; label given the commonplace rejection of large-scale hazardous facilities by the communities selected to site such projects
NRC   Michigan Natural Resources Commission
PURPA   Public Utilities Regulation and Policy Act of 1978
OCH   *Osceola County Herald*
PBB   Poliabrominated biphenyl, a complex organic chemical used as a fire retardant
PSC   Michigan Public Service Commission
RDF   Refuse-derived fuel
SECO   Steam and electric cogeneration
SORE   Committee to Save Our Forests and Environment
WFPP   Wood-fired power plant

Foreword

To some, the debate over siting wood-burning power plants, the subject of Fred Frankena's book, might well seem of limited importance. Yet, it focuses sharply on a type of issue that has occurred again and again over the past thirty years and has played a role of considerable significance in environmental politics. Most accounts of such issues have dealt primarily with the drama of the events and have not probed very deeply. Frankena's treatment, however, goes further to focus on the conflict over values and its relationship to population change, science, and technology as they apply to large-scale environmental intrusion in rural areas.

The conventional literature, for the most part, incorporates the political vision of the siting agencies, either private or public. Why, so the question runs, do communities seek to prevent justifiable siting actions that are "necessary" to foster economic objectives? The analytical context then becomes one of a conflict between desirable and undesirable public policies and the exploration of strategies as to how communities can be persuaded to accept what they should rightly agree to accept but do not.

Policy analysts have experienced considerable difficulty extending their vision beyond this limited context, to observe the interplay among factors embedded in the policy choices—factors such as values and value change, the dynamics of the course of scientific inquiry, and debate over the direction of technological innovation. Only if one can back off from the pressures of policy choice to the larger context of historical change in values, science, and technology can the limitations of the policy perspective be overcome. Because Frankena's study brings these factors into a more central role, it provides an opportunity to broaden the context and meaning of policy and policy history.

Values and Value Change

The most significant element in Frankena's analysis is the focus on values and value change. Much recent environmental understanding

15

has been concerned with process rather than substance, with how things are carried out rather than with objectives, with details about implementation rather than details about the origin and elaboration of values that underly objectives, with means rather than ends. By emphasizing values Frankena gets much closer to the heart of the historical meaning of environmental issues and environmental debate.

The continuing role of values and value conflict in environmental affairs is often obscured or even denied by environmental analysts and policy makers. Especially those in the regulated industries, administrative circles and professional institutions argue that there are no fundamental differences in values in environmental affairs. The fact that Congress has enacted much environmental legislation over the years, so the argument goes, reflects national agreement on basic environmental objectives. But the intensity of the continuing debates indicates otherwise. How else explain the innumerable conflicts that seem impossible to resolve? How else explain the language of basic values, "wood production fundamentalism" versus "environmental fundamentalism" that has come into play. How explain the notion often expressed that there are "religions" at odds here?

We must understand the rise of environmental affairs as a broad social development: the emergence of new values associated with a postmanufacturing society that came to the United States after World War II. The lives and values of people changed; what they wanted from public policy to enhance their standard of living now came to be discussed in terms of quality of life and to include a large dose of environmental quality. The change can be charted most precisely through the value changes associated with rising levels of education and which seemed to come largely at the "high school graduate" and "some college and BA degree" levels. These changes seem to be associated with issues such as health and smoking, changing size of family, and the changing role of women—and are also influenced heavily by education.

These value changes are closely related to changing consumption preferences over the course of time, an evolution from necessities in the nineteenth century to conveniences in the second quarter of the twentieth century, to amenities in the post–World War II years. Rising levels of education led to new aspirations, and rising levels of income and leisure enabled people to realize those aspirations. The changes have been identified and outlined most elaborately by those who chart consumer attitudes, the market analysts who now have gone beyond the traditional sociological variables to explore the psychological variables that sort out very different clusters of values even within similar socioeconomic levels.

It should be noted especially that these value changes did not come uniformly to all Americans at the same time. There are clear differences in the social context in which new values arose, were transmitted, and resisted. And that social context was often regional. Some regions of the nation were in the forefront in expressing new environmental values, and others were more reluctant to take them up. Suburban areas seemed to lead, and as they became the most vigorous location of population growth so also did environmental objectives gain more support. Rural areas often lagged behind, especially as their cultural context remained one of extractive and commodity production. But as rural areas and small towns became attractive places to live, their consumption and quality of life features often came to be more important than their commodity production features. It is this social context in which Frankena's wood-burning power plants take on larger significance.

In his descriptions of the organized opponents of the wood-burning power plants, Frankena adds to the evidence about the way in which new values, emerging in a broad demographic way, become transmitted from one place to another. In-migrants to rural areas, he convincingly argues, brought information and organizational skills with them that they had learned in more urbanized settings and through education. They provided more leadership, knowledge about access to information, and a willingness to confront the more powerful private institutions and public authorities than had traditional rural residents. This pattern Frankena describes is not unusual; it is repeated again and again in such cases of environmental intrusion.

Two aspects of this in-migration are particularly significant. One is its extent and character. Who and how many and over what period of time did they come? Migration to the rural countryside was observed continually during the 1970s and especially after the census of 1980. This was largely a matter of charting rates of population growth and comparing those in urban and rural areas. Hence, by the time of the 1980s, as these comparative rates changed and rural population no longer grew more rapidly than urban, observors argued that the process was over.

But, says Frankena, this may well be in error. First, there is rural out-migration as well as in-migration and the second may well have continued amid the first, thus bringing people with new values into rural areas even as rural population remained stable or declined. Second, the inflow of new values may not be a product only of population movement and change; cultural values, especially once in motion, are transmitted often by other means as well. Hence the transformation of values in rural communities could well come through more general processes of cultural diffusion rather than solely through migration.

It is also important to observe the relationship between the in-migrants and those already there. Rural reaction to environmental issues has been mixed. Rural people and their elected representatives generally have been skeptical of environmental objectives. Often they have been a major part of the environmental opposition. Hence they view zoning as an attempt by outsiders to control the free disposal and use of land in their communities.

But this reaction is often undermined by intrusions from afar in the form of actions by either private business or government in which local communities feel that they have become the victim of forces set in motion elsewhere that now impinge adversely on their daily affairs. In a wide number of cases, ranging over siting industrial plants and landfills, construction of dams and reservoirs, erection of electrical transmission lines, and extraction of coal and other minerals, long-established residents have become environmental activists, working in close tandem with residents more recently arrived. One can rightly describe this as a process in which new environmental values long dormant now are further elaborated and activated by a major community event.

In many rural communities the previously dominant commitment to commodity extraction or primary processing continues to shape community values. In these cases there is intense conflict between new and old, usually taking the form of bitter ideological con-troversy in which old and new values confront each other directly. Frankena's cases are somewhat different. For, as he emphasizes, the use of wood for fuel had long been a community practice and the innovation was the size and scope of that use. In fact, much of the controversy involved a conflict between older and newer fuel uses of wood; the new plants would jeopardize a long-used source of fuel for home heating. Local allies to those from afar who wished to intrude into Frankena's communities were relatively weak because of the degree to which the extractive economy was closely related to the consumer values of those who lived there.

Frankena's cases might well be contrasted with another type of case in which the intrusions and changes are smaller rather than larger in scale and come from within rather than from outside the community. Considerable environmental transformation occurs in less developed areas. Roads are widened and upgraded and so are bridges; farm land is subdivided, not necessarily leading to large condominiums but to large lots; buildings, driveways, and roads lead to a persistent but hardly observable decline in permeable surfaces; farmland is farmed more intensively up to the fence rows, reducing wildlife habitat. In such ways as these transformations that are massive in the long-run take place in such a way that they do not appear to be massive intrusions; they make their way with far less if not very little adverse response.

In the cases outlined by Frankena, large-scale, massive intrusions that boggle the minds of residents take place. The size and scale fits neatly with their own perception that the larger world out there, equally massive in size and scale, is an oppressive force against which they must defend themselves. That scale of intrusion, formulated and implemented by external private and public agents, instan- taneously shapes a massive conflict in perception and values. From that initial recognition, as Frankena shows, the long-run die is cast. Values and power become entwined as mutually reinforcing causal factors. It is well worth keeping in mind that many an environmental issue, not just those in rural areas, involve this close connection between values and power.

One aspect of this type of confrontation needs exploration beyond that emphasized by Frankena, namely the response of those in the industrial, administrative, and professional community to the reactions to their plans by those who live in rural areas. Sociological analysis of such "agents" is rarely undertaken. But they have their own peculiar perspectives, desires, values, and strategies—their own "culture" that needs to be brought into any satisfactory analysis of these environmental debates.

Over the years, experience by siting managers with local resistance has generated a distinctive series of reactions and ideas. Community opposition is, of course, unacceptable to the project proponents, whether private or public. There may be some willingness to modify the process so as to "listen" to the objectors and even to permit them to be participants in a limited sort of way. But the series of controversies, information about which is widely shared among development proponents, has shaped a distinctive point of view, a distinctive culture among those occupied with the task of siting.

One is the language of debate that identifies siting opponents as emotional obstructionists. Managers of development projects have long popularized the notion of NIMBY, "Not in My Backyard," with strongly negative connotations. It is not too much to say that the NIMBY phrase has become a negative symbol that arouses powerful emotions on the part of those who use the term and almost precludes careful and calm discussion. The bottom line is simply the idea that the projects, whether industrial developments or waste sites, "must" be put somewhere, and that those who object to siting them in their own communities are obstructionists. This context of thought and argument is now so firmly embedded that one can speak of an anti-NIMBY culture to be understood not so much in terms of logical argument but of intense and immediate emotional reaction.

A second aspect of the managerial culture is more strategic rather than ideological, the search for ways out in the form of communities that will not or cannot effectively oppose projects. This has taken several forms. One, and perhaps the most successful, is the willingness to provide massive economic benefits to communities to overcome their resistance. This strategy has long been used by private industry to make projects acceptable that would not otherwise be so, usually in the form of major tax contributions to the local community. Such economic benefits have now been fashioned, for example, as a part of the Pennsylvania state program to site a low-level radioactive waste depository. These cases sometimes work out with little fanfare if the community accepts the economic arrangement.

More frequently, however, the strategy has been to sort out communities that are politically vulnerable from those that are not, and to select the former in which to propose siting. Evidence about the location of waste sites in both urban and rural areas supports the argument that such a pattern is implicit in siting decisions. More recently, siting consultants have made explicit statements that siting proposals should not be made near middle-class or upper-middle-class communities. In a number of cases waste sites have been selected in poor, rural, and black communities, often in "unorganized" townships where resistance would be minimal. Such siting strategies have now been fully described by leaders of the "environmental justice" movement. It would seem plausible that Frankena's cases were among those in which siting managers gradually learned what kinds of communities were politically vulnerable to their objectives and what kinds were not.

Values, Science, and Technology

A major focus of Frankena's work is the role of science and technology in disputes over siting. It has been customary for many writers to pose this problem in terms of the validity or correctness of the scientific ideas under debate. In recent years, however, a more detached analysis has been taken up that understands technical controversy in terms of the professional and social context of those participating in the debate. Disputes are a result not of degrees of rationality or emotion but of different ways of looking at the world and choices about what in matters of knowledge should be emphasized over alternative approaches. This point of view establishes an analytical relationship between disputes over knowledge and disputes over public policy that is more deeply rooted in the human context and freer of normative choices made by the observer.

For many years matters of science and technology in both history and contemporary affairs have been analyzed much in the terms of how those fields of endeavor view themselves—the search for knowledge, done in an independent and fully objective way, followed by the application of that knowledge. The relationship between knowledge and application is thought of as either direct and unobstructed, carried out by those directly responsive to science and technology as objective forces or often bent and obstructed by popular emotion or by political forces on the part of those who are less knowledgeable.

An increasing body of research and literature, however, poses a far different context for understanding technical knowledge and its relationship to public policy. Conflicts within science and technology are often now understood more in terms of the internal dynamics of those professions, the values and conflicting directions within each discipline, and the differences in perspectives and methods between disciplines. These conflicts within the scientific and technical professions then are examined in their close connection with the larger conflicts in society and politics in fashioning public policy.

Frankena's analysis is in the spirit of this new approach to science and public policy. He emphasizes the connection between scientific controversy and policy controversy in the siting debate. It might be worthwhile, however, to place the entire matter in a broader context in which the beginning point is not the policy option that one identifies as desirable or undesirable, but the internal historical tendencies within those specialized professions themselves. To understand the relationship between divergent views within science and divergent views within policy one must begin with the internal evolution of science.

The key element in such an analysis is the growing pluralism of scientific expertise since World War II, a development in which environmental issues played a major role. Those issues brought into the scientific world a vast number of new realms of inquiry. Their exploration generated new specializations and along with many other new frontiers of science, helped to expand the range of individuals and groups seeking to shape the discovery and assessment of knowledge. In a considerable number of settings that brought together science and policy, debate shifted from a relatively small to a much larger number of experts and generated a new level of open competition and give-and-take among scientists and scientific ideas.

An early example of such a shift was in the field of atomic energy policy. Here an emphasis on new factors to be taken into account as consequences of siting—such as earthquake-prone sites or the aquatic effects of thermal discharges—led to new technical disciplines being brought into decision-making outside the narrower circle of atomic energy experts who had previously dominated the scene. In the health effects of lead, those emphasizing the potential adverse effect on industrial workers had earlier dominated thought about lead as primarily a problem in ingestion. Now the new interest in the effects of lead on child development sharply challenged older views and over the years completely transformed thought about lead science with attendant consequences for lead policy. In our own day the health consequences of toxic chemicals, once confined primarily to cancer, are now being challenged by those who emphasize a wide range of other effects—genetic, reproductive, fetal, pulmonary, and on immune systems.

Cases such as these build up a set of generalizations about the direction of historical development in such matters. Earlier the context was one of scientific and technical expertise within a relatively closed setting of governmental agencies and professional specialists associated with private industry and academic institutions who sought to keep scientific and technical debate relatively free from larger public influence. As time went on, however, the context became more open, in which a wider range of specialists participated and public debate replaced private discussions, often with media participation. Such was Frankena's case of wood-burning power plants. Several elements in the process deserve a bit more attention.

The most fundamental aspect of this shift from a more closed to a more open context of scientific and technical debate was the role of specialization. Within science and technology themselves, specialization leads to varied scientific "publics" that enhance diversity of opinion. Specialists create new institutional worlds often isolated from one other. The ideology of science assumes that the scientific method generates a high level of agreement. But, on the contrary, enhanced scientific specialization enhances disagreement. This discordant tendency arising within the culture of science itself is often obscured by the tendencies of scientific bodies, such as the National Academy of Science, to avoid minority and dissenting views in its reports.

But the divergence of views within science was more powerful than these attempts to impose unanimity on diversity, and in the 1970s many dissenting scientific opinions emerged on the public scene. Its roots lay in disciplinary variety. As specialization took place so did the conviction that one's distinctive way of looking at the world should receive greater recognition in the larger context of debate and decision-making. Hence specialization in science and technology carried its own inherent tendency toward centrifugal expression and debate.

This interdisciplinary process was closely intertwined with institutional competition to stimulate variety in opinion and ideas and hence debate and disagreement. One source of this competition came from academic institutional support for science and technology. Academic institutions compete for recognition and in the process often seek to advance one line of inquiry over another. In some countries the centralized organization of research institutions tends to keep these centrifugal tendencies in check, but in the United States the vast number of such institutions, each closely connected with self-images of state and regional prestige, foster assertions of the importance of new research in competition with old, of unconventional ideas amid the conventional. Institutional competition, therefore, becomes a major force for pluralism in scientific and technical debate.

Institutional variety also marked the role of federal agencies in enhancing pluralism in scientific debate. The years after World War II not only witnessed the rise of the overarching scientific branches of the federal governments, such as the National Academy of Sciences and the National Science Foundation, but also a wide range of scientific enterprise in administrative agencies. In matters of health, the Centers for Disease Control and the National Toxicology Program became independent centers of initiative with respect to environmental health effects. The Geological Survey contributed its own on-going water quality measurements and seismological assessments to policy issues arising in the Environmental Protection Agency (EPA) and the Atomic Energy Agency (AEA). The U.S. Fish and Wildlife Service generated its own science about "effects" of a wide range of developments on aquatic life. These tendencies in federal agencies were reflected in and enhanced by the requirement in the National Environmental Policy Act that environmental assessments be "interdisciplinary" and "wide-ranging," a requirement that the federal courts underscored as essential for sound administrative procedure.

State governments have also become important in contributing to the context of scientific pluralism. At one time the federal government was all-powerful in scientific capacity and authority. Especially in environmental affairs, such as environmental health and ecological knowledge, states were heavily dependent on federal agencies for technical knowledge and expertise. With time, however, as states became more affluent, their scientific and technical capabilities increased. California, for example, developed environmental capabilities about air pollution and pesticides that could often challenge federal scientists. In the first round of the dispute over the health effects of the pesticide alar, both Massachusetts and New York developed their own toxicological analyses that enabled them to challenge the scientific conclusions of federal agencies. In the 1980s a coalition of eight northeastern states pooled their scientific and technical resources, often in cooperation with California, to take up air pollution issues, thus extending the state role in scientific and technical pluralism. Such tendencies often led to scientific conclusions from the state level that were at variance with the federal, thus extending participation in scientific debate in public policy.

Citizen organizations—and this is the focal point of Frankena's argument—played an important role in this tendency toward technical pluralism. From the earliest days of modern environmental affairs in the 1960s, citizens concerned with environmental problems reached out to technical specialists. They well understood the role of technical knowledge in policy decision and recognized that if they were to challenge an action they considered to be environmentally detrimental they would have to challenge the technical basis of that action. This was often not difficult to do since many proposals were accompanied by a mere assertion of technical fact rather than an ample demonstration of it and citizens were quite able to bring technical expertise to bear in exposing those limitations. They were especially able to inject into the scientific debate the insight from a disciplinary viewpoint different from those used in a competing analysis.

Thus it was, for example, that in 1960s, when the Pacific Gas and Electric Company sought to site a nuclear reactor at Bodega Bay, the citizen's group that opposed it found a geological consultant who was able to raise serious questions about the geological safety of the site. The testimony of that expert, in turn, prompted the Secretary of the Interior, Stewart Udall, to bring the U.S. Geological Survey into the fray and because the Atomic Energy Commission (AEC) had almost no geological expertise at its disposal, the debate over the adequacy of the site soon moved from the close relationship between the utility and the AEC to the larger realm of public discussion. Or, a few years later the U.S. Fish and Wildlife Service was able to bring significant arguments into atomic energy decisions about the detrimental effects of thermal discharges on aquatic life. This case, in fact, played a major role in shaping the requirement that environmental impact analysis be fully interdiscipinary.

In this and many other such cases, the role of citizen organizations reflects a major development in American private and public life— the education of a vast number of individuals who while not technical experts themselves, are knowledgeable about the role of such expertise and where it can be found, and have some keen sense of methods required to ferret out and evaluate knowledge. Common to a wide range of citizen actions is the participation in such groups of people who have had several years of college education, who have undertaken searches as academic projects under the guidance of college and university instructors, and whose investigative skills, acquired in such training, they now apply to their current environmental circumstances. There are also major cases in which people with only a high school education, because of some incident in their community, became self-educated to such an extent that they could challenge the experts associated with public and private agencies. Adult environmental education of this kind has played a major role in modern environmental affairs.

Such ventures, moreover, do not remain simply individual inquiries. For soon the citizen "expert" begins to participate in a world of expertise involving both academic specialists and other citizens who are becoming equally knowledgeable about the subject with which they are concerned. Knowledge becomes shared within networks of amateur and professional specialists. In some cases, a sufficient number of people become interested in the subject so that an organization is formed to become a central clearing house for technical information. Here technically trained people are hired to keep in touch with on-going research, follow the journals, establish continuing contacts with professional specialists in universities and government, and serve as major vehicles for scientific and technical transfer—not only among themselves but to decision-markers in government. In some cases such activities have given rise to regularly published summaries of research to the extent that they are of distinctive value to specialized researchers in academia, government, and private industry.

The enhancement of pluralism in scientific debate and hence in the role of science in public policy has been met with misgivings on the part of many of the most influential leaders in scientific institutions. They are often prone to speak of the resulting controversies in terms of reason versus emotion and good versus bad science. These reactions are understandable as sociological and political responses to innovation in the realm of science and public policy. They are also understandable in terms of the perennial conflicts between old and new in historical change. However, by themselves they are not a useful guide to careful social analysis. New ways of comprehending these disputes are essential in order to assess accurately their role in public affairs.

There are many worlds of scientific and technical environmental networks linking specialists in academic institutions, and state and federal governments with those in citizen groups. Frankena's work is a particular case example. It is not by any means the first of such studies and his references can guide the reader to many more. But it provides an excellent account of the way in which scientific and technical information now plays a role in a far more pluralistic universe of expertise than was the case a quarter of a century ago. It is, thus, a case study not only in the history of changing values but also the history of the evolution of science as a public process.

Values and Science

These two aspects of *Strategies of Expertise in Technical Controversies*—values and science—deserve wide attention, for the heavy policy preference emphasis of most environmental writing has all but ignored both the evolution of environmental values as an historical process and the sociological and political context of expertise.

Value change as a broad-based demographic process seems to fall completely outside the perspective of most environmental analysts. At the same time, they are so absorbed in the task of persuading their publics of the "right" solution to a problem that they instinctively shy away from a more detached observation of choices made by experts.

We who have been professional observers of environmental affairs over the past several decades have often neglected the real world of human values and scientific debate. We have been so preoccupied with following the intricacies of environmental policy formation and implementation that we have neglected the context of historical and

sociological understanding. Frankena's study provides an opportunity to set the analysis of environmental public affairs off in quite
a different direction and one that would greatly enhance that
understanding.

SAMUEL P. HAYS
Pittsburgh, PA
1 October 1990

Acknowledgments

This book is the product of a long evolution involving the help of many people. To begin at the beginning, J. Allan Beegle provided essential financial support on a related project in Osceola County. The compatibility between the study of the population migration turnaround he directed and my research on the Hersey controversy has contributed directly to the ideas in this book. Daverman and Associates of Grand Rapids, Michigan and the *Osceola County Herald* based in Reed City, Michigan generously loaned materials analyzed in this study. Michael McClelland of the Michigan Public Service Commission put me in touch with information and contacts relevant to recent efforts to develop wood for energy in Michigan and elsewhere. Pacific Energy of Commerce, California, in particular Tony Henrich, was most cooperative in rendering a number of photographs of wood harvesting operations and wood energy facilities in the West. Among opponents to the Hersey project, Marco Menezes was indispensable. He provided materials published by the Committee for the Rational Use of Our Forests, responded to my questions on numerous occasions, and thoughtfully critiqued some written accounts of the case history.

I would be remiss if I did not acknowledge the help of my doctoral dissertation committee—Marilyn Aronoff, J. Allan Beegle, Peter Kakela, and my major professor, Harry Perlstadt. I owe a special debt of gratitude to Harry for so capably directing and evaluating my work.

A number of individuals and experiences encountered after the original case study have left their mark. Two conferences—The First International Conference on Consumer Behavior and Energy Policy held at Nordwijkerhout, Netherlands in September 1982 and the First National Symposium on Social Science in Resource Management convened at Corvallis, Oregon in May 1986—brought me into contact with researchers and studies that have contributed significantly to this work. Anonymous reviews by several scholars, especially the reviewer for Lehigh University Press, have served to greatly improve presentation of this study.

I would also like to express my deep appreciation for the work of two scholars—Dorothy Nelkin and Samuel P. Hays. They are often cited in this book and for good reason.

Last but not least, I wish to thank the many libraries and librarians that have assisted me. My wife, Joann—to whom this book is dedicated—is my favorite librarian. Throughout its writing she has been an able assistant, a constant source of encouragement, and a model of patience for anyone who must abide the trials of an author.

Notwithstanding these diverse contributions and influences, I claim full responsibility for all errors and omissions. The views expressed are mine, with generous citation to the many and varied sources for this synthesis. My interpretations and conclusions are designed both to stimulate further thought and research on the social study of expertise and to indicate my own values so that the reader may better fathom my reasoning and judge my biases.

Introduction

Experts have become so much a part of our lives that we take them for granted. A common image of experts is men or women surrounded by books in a dingy office contemplating weighty scientific theories, or clad in white suits while poised over the latest experimental technology. They are viewed as technicians who are oblivious to politics and who are only interested in the pursuit or application of scientific truth.

As is often the case, the popular notion does not accord with reality. Scientists and technicians are seldom called upon to find or apply scientific facts to the exclusion of social value judgments. Studies of technical controversies, particularly environmental controversies, have demonstrated that experts and expertise are often employed to mask political choices. Studies have also revealed that, in response, counter expertise is utilized in an attempt to open the decision-making process. Indeed, experts and expertise have become a political resource, not simply the wellspring of scientific truth or new technology. This is the sum and substance of the strategies of expertise. The process is well established in postindustrial society. The ensuing pages constitute an exploration of this phenomenon.

Large-scale development of wood for electric power in the United States serves as the vehicle for this study. At its core lies a particular case—the proposal to build a twenty-five megawatt (MW) wood-fired power plant (WFPP) at Hersey, Michigan. The book moves beyond this case to a comparative review of subsequent wood energy controversies, including WFPP siting conflicts at Indian River, Michigan; Quincy and Westwood, California; and Burlington, Vermont.[1]

This is not another academic treatment of a narrowly conceived problem that is of interest to only a handful of researchers. Rather, this is a calculated attempt to move in the opposite direction—to understand the community in action rather than as the nebulous sum of continually subdivided and analyzed parts. It is assumed here that a holistic framework is required to study the strategies of expertise. The problem is at once social, political, and environmental. Strategies of expertise find their origin in social change, environ-

31

mentalism, the energy problem, and the politics of science and technology. To inquire by breaking the problem down into the specialties and subspecialties of social science—and seeking to quantify and refine knowledge within these subdivisions—is a flawed methodology for understanding technological controversy. Controversy involves the whole community and its environment. It is a mix of social and natural that defies the academic division of labor. This is not to say that the division of labor is an inefficient way to gather information, merely that somehow aggregating this information does not necessarily illuminate the overall problem. Understanding controversy and what it implies for resource development in a democratic society requires such a grasp of the whole.

This book observes social behavior and draws upon research for a wide range of topics. Diverse disciplinary threads are thereby woven into a tapestry of intricate detail. With the benefit of hindsight, it is difficult to envision how an adequate account might have emerged from a more narrowly conceived approach.

An analogy can help to further justify the perspective taken in this study. On a drive through the countryside a layman looks at the roadside and sees a field swept by the wind. The scene is rendered visible by sunlight. Grass and shrubs are apparent and perhaps some wildlife. To the layman, it is not at all obvious what the relationship is between these things. The observer may appreciate the beauty of the sight he or she beholds, but probably does not understand the interrelatedness of the things observed. In the absence of curiosity, any deeper significance of what is seen will remain a mystery.

In contrast, an ecologist presented with the same scene perceives the powers of nature at work. The energy of the sun is seen as the supreme force for both the physical environment and the sustenance of life. Energy is transmitted through the ecosystem by trophic relationships among and between the plants and animals that live there. Although the web of life is not grasped by the ordinary observer, the ecologist is educated to perceive it and its telltale signs. For example, grass and shrubs derive their energy from the sun through photosynthesis. Insects feed on these plants and in turn become food for birds. Together the physical forces of sunlight, wind, and seasonal change affect the way organisms adapt to their environment. Appreciating the ecosystem in these terms requires the trained eye of the ecologist. Once achieved, this understanding forever changes the way the scene is observed and understood.

An encounter with technical controversy is not unlike this view of the countryside. The layman sees technical controversy as simply a fight between two groups or two opposing interests. One group wants one thing and the opponents want the opposite. The dispute may involve protest, lawyers, scientists, and government agencies. When it is all over, one side usually wins and the other side loses. In this it is no different from an athletic contest, but there is much more to controversy, just as the tangible parts of the countryside in the above example are not all there is to an ecosystem. The dynamics of controversy are there to be studied and "seen" by the interested social scientist. Guided by previous studies of conflict, he or she observes the effects of ideology, values, expertise, and social change in creating and sustaining controversy. This is a view quite different from the layman's.

If informed choices about environmental issues can be derived from understanding ecosystem dynamics, then it goes without saying that understanding controversy would greatly benefit public policy in a democracy. And just as citizens might better participate in decisions about wildlife issues by learning some ecology, they come to understand the functions of social conflict by learning about technical controversy and its alternatives. Included in the class of laymen are decision makers in government who are often stymied by controversy, yet who do not understand its dynamics sufficiently to steer clear of its perils. If this study of wood energy development prompts some decision makers to reflect on their role, on the way that they utilize experts and expertise, and on the extent to which they involve the public, then its purpose will have been well served. By the same token, if the study shows concerned citizens how the process of technical decision making works and what they must do to express their values in the process, then it will also have proven worthwhile. There is an unmistakable symmetry in the strategies of expertise.

Two complementary problems occupy this study. On one hand, the study seeks to better understand technological controversy and the role of experts and expertise in controversy. Scholarly questions more of interest to the academic community are the central focus of this effort. On the other, the practical consequences of this knowledge become crucial in an era of rapid technological change. Citizens, government, and the technical experts themselves all have an abiding practical interest in understanding controversy. Therefore, the book addresses issues of policy and administration connected with this social research.

The dichotomy of interest applies both to the social study of experts and expertise and to the social problem of the development of wood for electric power. Policy implications are discussed for technical controversies in general and wood energy development controversies in particular. Consequently findings should appeal to an especially diverse group of scholars and practitioners in the specialties of environmental studies, forestry, planning, political science, public administration, resource development, and sociology. Prominent issues more or less relevant to each of these fields include citizen participation in decision making about resource development, natural resource/environmental conflict, the social and political impact of the population migration turnaround, ideology in energy resource development, and the politics of experts and expertise.

Because this is a broadly based social study of wood energy development, an introduction to this subject is required as a prelude to treatment of relevant concepts of expertise. Wood is an ancient energy source that experienced renewed interest in the wake of the energy crisis beginning in the 1970s. Residential woodburning, for example, has increased dramatically since the oil embargo in 1973. The concept of using wood to produce electric power has also been actively pursued during this period. Interest in wood as an alternative source of energy persists despite the substantial decline in fossil fuel prices during the mid-1980s. A number of projects have been proposed and a few are now operating. Yet large-scale development project have not achieved the success envisioned by those promoting the concept. Why a renewable source of energy should prove so difficult for developers is an important facet of this book. It critically examines public controversies incited by proposals to develop WFPPs in Michigan and elsewhere. These controversies clearly demonstrate continued public aversion to large-scale, environmentally hazardous technologies—a dislike that extends even to renewable resource development.

The context of proposals to develop wood for electric power is both complex and dynamic. It is fraught with emergent social changes for which history cannot serve as a guide. During the 1970s the environmental problem and the social movement it inspired focused attention on technology as a social problem. Controversy has been almost synonymous with the siting of large-scale energy production facilities, in particular nuclear and coal-fired power plants. A series of supply crises have revealed an enduring energy problem. The fact that fossil fuels are nonrenewable resources that are being depleted at an ever increasing rate has created an unprecedented challenge for energy policy. New views of energy development, based on renewable sources, have stimulated an energy policy debate. Beginning in the 1970s many rural areas experienced an unexpected return to population growth. This turnaround, as it is called, has increased pluralism in these areas. Local politics have been altered as newcomers aspire to influence decision making in their adopted communities. This brings up perhaps the most crucial change of all—the rising tide of citizen participation in decisions that all too often have been defined as belonging to "the experts." Citizen groups are forcing administrative agencies to be responsive. Where mechanisms are lacking for the public to be heard and heeded, the politics of protest have often proved effective in opposing controversial decisions. But there is more to it, as this account of the strategies of expertise demonstrates. Emergent social factors are intimately tied to controversies surrounding proposals to build WFPPs. Studying the behavior of communities faced with such controversial proposals leads inexorably to consideration of these macrosocial changes.

The milieu of resource and sociological factors associated with energy development was prominent in the major case studied here. The Hersey, Michigan controversy is the classic case of opposition to WFPPs. It began in 1978 during the site selection process and ended in withdrawal of the proposal in 1980. How did a controversy arise over such a seemingly benign development? Why did it follow the general pattern for public opposition to large-scale technologies in pluralistic communities? What does this suggest about resource development in general and the development of renewable resources in particular? Answers to these questions for the Hersey case provide the baseline for observing and evaluating subsequent development efforts.

After the defeat at Hersey a period of stalemate in northern Michigan was followed by other proposed projects. The 15-MW WFPP planned for Indian River, Michigan became controversial in 1983 with similar results.

Two WFPP controversies in California and one in Vermont are also examined in detail. Two 11.5-MW plants were proposed for the northern California communities of Quincy and Westwood. The world's largest wood-burning plant, a 50-MW facility, was constructed at Burlington, Vermont.

Of these five proposed developments, three were defeated and two were built. Controversy enveloped all of them. The most controversial has been the Burlington facility where adverse impacts seem to have matched the scale of the plant.

The essential method for coming to terms with the strategies of expertise as manifested in environmental controversies is the case study. Case studies are especially commendable in circumstances where, as in this one, the challenge of complexity and interaction effect is great.[2] The design used is based on Yin's conceptual framework for case studies.[3] A comparative structure is employed repeating the same case a number of times. In each repetition it is examined from a different perspective or analyzed with new data and information. Additional cases of the same type are also considered, providing an opportunity to compare patterns of behavior and to generalize findings of the baseline study. Within this structure, the general purpose is to describe and explain the strategies of expertise and their contribution to wood energy development controversy.

At the same time, this is a revelatory case study. Wood energy development promulgated by the energy crises of the 1970s is an emergent response with unique features and it is important to understand this new problem. Nevertheless, a study of this case can also contribute to general knowledge of environmental and energy facility siting controversies.

Several conspicuous aspects of controversy are briefly summarized to set the stage for this study of wood-electric power controversy. Social conflict, the interaction of values and ideology, and resistance to technical change, respectively, are each themes in this study.

Controversy is social conflict. A useful definition for social conflict is a struggle over values or claims to status, power, and scarce resources, in which the aims of the conflicting parties are to gain the desired values as well as neutralize, injure, or eliminate their rivals. As such, controversies involve the distribution of scarce values or goods. In siting controversies environmental values are often at issue. Development values come into conflict with local environmental values. Such matters are not objectively determined—rather, the actors in the conflict define the situation. Analysis of controversy, therefore, must consider the values and ideologies that give rise to the conflict.[4]

Values are undoubtedly the most significant underlying cause of technical controversy. Individuals act upon their values; values are embodied in ideology, which in turn shapes the course of conflict. Based upon their values, individuals form groups. Ideology, like expertise, serves as a strategic tool. It is the means by which activists recruit and mobilize followers around a vision of reality and a set of broad social goals.[5] This case study seeks to contribute to understanding technical controversies by unravelling the evaluational bases of their competing interests in terms of values and ideology.[6]

The issues in technical controversies are increasingly seen as involving choices between competing sets of values and not just between technical alternatives. The tension between the technical-scientific and the normative invariably gives rise to controversy. With this is mind, it is important to note at the outset that technical knowledge is utilized as a tool and exploited by divergent interests to establish their claims.[7] Furthermore, where values conflict and experts are found to disagree, a political solution is inevitably required; policymakers have little option but to fall back on ideology, interest, group pressure, and the like for making decisions.[8] Despite the apparent importance of this thesis, it is a neglected subject in the social sciences. One motivation for this book is the desire to remedy this neglect.

Resistance to technical change implicit in siting disputes arises at least in part from concern for local autonomy—a response to the powerlessness inherent in the pervasive influence of expertise. Resistance is also due to concern about global questions of value—a reaction to the abrogation of values epitomized by reductionism in science.[9] Both fundamental aspects of resistance to technical change are evident in this study.

Data and information for this study are largely derived from documentary sources including transcripts of public hearings, local environmental reports, newspaper and magazine accounts, and records of meetings. Interviews with government officials, business professionals, and citizens were utilized to fill in the details, enhance the validity of the information gathered, and more fully explore the social complexity of technical controversy.

The book is divided into three parts. Part I: Experts, Expertise, and Social Change provides background information and a review of literature pertinent to studying wood energy development controversies. Part II: Wood Energy Controversies details the cases. It discusses the nature of wood energy development and comparatively treats wood energy controversies. Finally, it assesses wood energy policy in terms of what has been revealed by the study of related controversies. The final four chapters comprise Part III: A Social Study of Woods Energy Controversy. They contain a reanalysis of the cases and general conclusions about the study of controversy and the tension between technocratic and democratic decision making. The chapters presented within this three part framework are briefly summarized below.

Chapter 1 defines and analyzes expertise and its political uses. The strategic use of expertise as a resource in technical controversies is emphasized. Prominent social changes that have given rise to and sustained wood energy development controversies are treated in chapter 2. Topics covered include environmental awareness (using the Michigan PBB contamination as a salient example), power plant siting and the so-called NIMBY ("not-in-my-backyard") syndrome, the population migration turnaround, and the sometimes opposing ideologies of community economic development and local autonomy.

Wood energy concepts are presented in chapter 3 as a basis for understanding the technical debate studied in subsequent chapters. The next four chapters present the WFPP controversies. Chapter 4 discusses the impetus for the resurgence of interest in wood energy in Michigan, the early progress of the proposal to build a demonstrational WFPP, and then summarizes the controversy resulting from the decision to site the plant at Hersey, Michigan. Wood energy controversies in Michigan, California, and Vermont are examined in chapter 5. A parallel controversy at Indian River, Michigan offers additional support for findings of the Hersey study. Controversies with contrasting results in northern California and Burlington, Vermont provide further insights. The California study is then compared with the Michigan cases of WFPP development. The comparison focuses on explanation for the conflicts observed. Chapter 5 concludes with a brief summary of WFPP development and some attendant controversies in the Northeast United States. Some implications of the study for wood energy policy and political institutions occupy the final chapter in Part II. Chapter 6 illustrates the policy dilemmas of wood energy development. It focuses on the State of Michigan. The *Michigan Wood Energy Plan* is summarized and evaluated in terms of the strategies of expertise.[10] This policy evaluation relies on three significant issues emerging from the study—technological scale, local autonomy, and citizen participation.

The wood energy controversies are further analyzed in the next two chapters. Chapter 7 looks at controversy in terms of facts, values, and ideology. Content of the public debate in the Hersey case is analyzed to assess the relative impact of facts and values. This information is used to illuminate the study of experts and expertise in chapter 8.

The final two chapters generalize findings to the problem of wood energy development and more broadly to environmental controversies. They are oriented, respectively, to the study of controversy as an academic enterprise and to the practical matters of policy and practice for a democratic society. Chapter 9 seeks to extend our understanding of controversy, drawing upon the analyses and cases in the foregoing chapters. Finally, chapter 10 offers implications of the study for the technocratic and democratic styles of decision making, providing general conclusions about the strategies of experts and expertise in technical controversies.

The fundamental issue in this study is the involvement of citizens in decisions that affect their community and its environmental values. The theory and practice of citizen involvement necessarily requires an appreciation of the strategies of expertise. It is argued that the strategies of expertise are neither a threat to the integrity of science nor the death knell for democratic values. Disagreements among experts and concomitant technical controversy may be healthy activities in a democratic society, at least to the degree values are ignored or concealed by technocratic decision making.

A digression regarding science and values should help set the tone for this book. The notion persists that social science should steer clear of values even to the point of excluding them from study. In this manner social scientists continue to subscribe to the positivist ideology of the natural sciences. Yet this tendency conflicts with the emergence of social studies of science and technology which has established the essential place of values and value judgments in scientific research and application. Those closest to the problem have come to adopt a different perspective. Debates about "value-free" science are heard less often these days. Although it may serve as an ideal for natural scientists who generally abjure social issues, the ideology of logical positivism severely limits the comprehension of the social scientist. Experts and expertise, to cite the salient example, are inextricably tied up with values.[11] Social researchers must seek to illuminate values as well as facts.The subject matter requires it.

By the same token, crucial value judgments all too often lay hidden behind the veil of scientific objectivity. In my view, it is incumbent on researchers to give an indication of their values. Readers who are aware of a scholar's values are better able to determine how they may have influenced presentation of the facts. Accordingly, I should like to give some indication of my values at the outset. The objectives of this book are to find the truth about technical controversies and to seek ways to improve democratic decision making. Both endeavors reflect values, and both require consideration of values. Knowing this should help the reader judge the success of my efforts.

I scrupulously observe, analyze, and report facts in this study through the prism of foregoing social science. At every turn the social situation presents values—the values in the controversy itself and the values prospectively affected by the findings of the study. I have tried to give appropriate weight to the values and the facts of wood energy controversy. I sincerely hope that this study is, as a result, more truthful as well as more practical.

Strategies of Expertise
in Technical Controversies

Part I
Experts, Expertise, and Social Change

Experimentation and Social Change

1

The Politics of Expertise

The relationship between the expert and the public is becoming
more crucial, more intimate, and more complex than anything
that the theory of democratic government could possibly have
anticipated.... We neglect this relationship, in theory or
practice, only to our peril.[1]

Understanding the role of technical experts and technical information in public controversies becomes a more important research
question as the number and diversity of such controversies increases.
For this reason, the neglect of this subject in the social sciences has
become an acute problem.[2] Technological innovations are created
with little concern for their social impact or public reception. As a
consequence of the rising anxiety over the hazards and risks of such
technologies as recombinant DNA and radioactive waste disposal,
technical expertise is increasingly used to challenge as well as to
promote controversial decisions. Science and technology have come
to serve as political resource exploited by various interests to justify
their claims. This trend has eroded the success of proponents in
depoliticizing political controversy by mobilizing technical experts.
Put differently, promoters of technical innovations have sought to
institutionalize technical expertise in order to cope with controversy.
This approach often fails because the public is interested in matters
of value as well as matters of fact.

Technology creates problems that are both technical and
nontechnical. Narrow definition of issues as "technical" has been
central to public hostility toward technical experts and public
apprehension about technology. Science and technology cannot
accomplish every task. "Scientific methods were not, are not, and
cannot be intended as a substitute for public participation."[3] Those
calling for increased public participation as a remedy are too
numerous to cite here. Suffice it to say that short of full public
participation lies the use of counterexpertise.[4] Citizen opposition

45

groups have learned that if expertise can serve as a weapon for social manipulation in the name of rationality, then it can be used to create conflict to serve their ends.[5]

The 1970s have been labeled the "golden age of environmental conflict."[6] Part and parcel of this conflict has been the use of experts by citizens' groups to oppose cadres of well paid institutional experts. Volunteer activists from within the scientific community were on the vanguard in early environmental conflicts. The development of public interest organizations concerned with the applications of science and technology provided niches for many such activists. But from the inception of the environmental movement, committed, well educated individuals have volunteered their services to oppose controversial decisions. Pursuant to the help of such volunteers, the year 1970 proved to be a watershed in the ability of environmental groups to play an effective role in the politics of technical expertise. Prominent scientists had volunteered to oppose such controversial technologies as nuclear weapons well before that year. However, 1970 marks the beginning of widespread volunteering by experts in a diversity of controversies. It is in this context that the social role of the volunteer expert is construed as emergent.[7] The historian Samuel P. Hays describes the change as follows:

> In each environmental episode there were usually a number of people who had learned skills of investigative research in college. Many citizen groups could enlist the aid of technical expertise outside their ranks. And often some within their organizations were biologists, chemists, engineers, epidemiologists, economists, and lawyers.[8]

This chapter begins with a synthesis of studies and statements relevant to understanding the role and impact of expertise. It provides a conceptual framework for the institutional role of expert, the corresponding impetus for the role of voluntary technical expert, and the behavior of both in a milieu of facts and values.

Experts and Expertise as a Political Resource

The concept of expert is social. It is usually associated with controversies because they involve complex problems that must be interpreted. The expert may be employed by one party to the controversy or may be acting as a committed volunteer. In any event, expertise involves special skill or knowledge in a particular field or specialized discipline.[9] Although those filling the role of expert are expected to make an honest presentation, they are understood to be liable to systematic influence in coming to opinions on technical matters. The notion that expertise can be biased is well appreciated by judicial institutions which have a long history of dealing with expert witnesses.[10] The expert is now called upon to answer to the challenge of laymen in public controversies as well as in courtroom litigation.

Expertise has come to serve as a political resource.[11] Government administrative authorities have sought to depoliticize political controversy by employing expertise. Controlling the way expertise is institutionalized has enabled authorities to manipulate the content, use, and power of technical expertise. Scientists and technicians have eagerly acquiesced given their tendency to interpret and define the technical aspects of a problem as though they were the whole problem. Twain's remark—that if your only tool is a hammer, all problems look like nails—rings true for experts. Controversial social and political issues have often been dealt with by limiting problem definition and assumptions. By such means scientists and technicians have been subjected to the requirement that their conclusions be politically useful.[12]

Technical expertise is a crucial political resource in controversies because access to knowledge and the resulting ability to question the data and information used to legitimate decisions is an essential basis for power and influence. Making the debate technical restricts participation to those who are capable of technical argumentation. Dissenters soon realize that expertise can be bought. Citizen opposition groups have found it both necessary and fruitful to use this tool in opposing administrative decisions. The complex data supporting a particular development cannot effectively be challenged without resorting to counter expertise. Without some command of technical information, opponents are powerless in the world of administrative politics. Technical expertise is now understood to be the only way to gain a foothold in controversial decisions whatever the underlying political or moral objections.[13]

But what happens when expertise is marshalled by both sides in a controversy, that is, when it becomes a weapon in both of their arsenals? When the experts disagree, the result is usually further polarization of the conflict which intensifies the controversy and often ends in the defeat of the proposed project.[14] The scientists and technicians in the dispute are revealed to be fallible, demystifying their special expertise and calling attention to nontechnical and political assumptions that influence technical advice. Disputes between experts invariably lead to public confusion and in the end push public opinion toward opposition rather than acceptance.[15]

A conceptual framework for the study of expertise requires acknowledgment of the social roles that are available—a continuum ranging from *consultant* on the one hand to *adversary* on the other, with the *expert* and the *advocate* lying in between.[16] Figure 1 presents a model for understanding the range of possible roles for actors in a controversy. The set of roles should be viewed as a continuum in which the distinction between stereotypical roles is not well defined (see note accompanying figure 1). The corresponding behavioral tendencies appear below the model in the figure. Changing roles from *consultant* to *expert* to *advocate* to *adversary* at the opposite pole of the continuum involves moving from an apolitical role to a political role, from primary concern with facts to primary concern with values, from determining the scientific truth to winning in the political arena, and from application of scientific method to legal and bureaucratic maneuvering. Although most university trained professionals see themselves in the role of consultant, in fact the role each assumes usually resembles that of expert or advocate for the interests they serve.

The interpretation of any scientific/technical statement is subject to the influence accorded the roles that are accepted by the actors and audience in a dispute. It is by no means certain that special credentials cast an actor in the role of the dispassionate consultant. One observer cautions that "those who are assessing statements which purport to be scientific in such situations must judge the person well."[17] Whether one acts as an adversary, advocate, consultant, or expert depends on how s(he) behaves in the controversy. Yet even the most objective scientific advice can appear to have been politically motivated when it is carefully exploited. The most well intentioned expert can be exploited to support political positions.[18] These roles are not defined a *priori* or by the interest groups in the debate. The controversy defines roles as surely as it unmasks values, leaving it to the unbiased social observer to sort them out.

The term "expert" must be understood to have a general and a specific meaning in order to properly interpret figure 1. Generally, all of the roles available to actors providing specialized knowledge reflect expertise. As such, each role carries the implication that the actor is "the expert," or at least possesses specialized knowledge. Notwithstanding the general meaning of expert, there is a specific role which is best described by use of the term. It lies between the unshakeable consultant on the one hand and the committed advocate

---

FIGURE 1

A Continuum of Roles for Social Actors Providing Technical Knowledge:
Consultant, Expert, Advocate, Adversary*

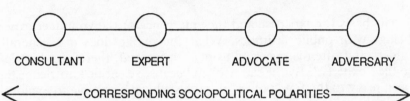

CONSULTANT   EXPERT   ADVOCATE   ADVERSARY

⟵——————CORRESPONDING SOCIOPOLITICAL POLARITIES——————⟶

apolitical ....political

factual orientation ....values-oriented

goal to determine truth ....goal to succeed in
                                                                political arena

objective pursuit of...subjective use of
knowledge   knowledge

passive role...active role

scientific method..legal/bureaucratic
                                                                maneuvering

---

* The concepts in this model are by nature dialectical (see Nicholas Georgescu-Roegen,
   The Entropy Law and the Economic Process). Dialectical concepts are not discretely
   distinct. In the words of Georgescu-Roegen, they are "surrounded by a penumbra within
   which they overlap with their opposites." Although it is clear that an expert is different
   from a consultant or adversary, the boundaries are not obvious. The lines between
   stereotypical roles and the notion of a continuum are used in this figure to represent the
   dialectical character of these concepts.

---

on the other. As it happens, those who fill the role in question are
commonly referred to as experts. In the judicial arena "expert
witness" is the exclusive category.[19] As will soon be revealed in
connection with wood energy development, experts tend to move to
the right of the continuum described in figure 1 when controversy

arises. That "experts" in a general sense do so does not diminish the functional role of expert in the more restricted sense of the term.

The Impetus for Voluntary Expertise

The basis for expertise and the emergence of the volunteer expert beings with public attitudes. What the public does not generally understand often is regarded as mysterious and, therefore, best left to the "experts." By the same token because scientific knowledge has been privy to the few, powerful interests have been able to exploit it as a political tool. Opponents in technical controversies usually have little political power. The spread of knowledge via the greatly expanded group of college graduates is producing a countervailing effect by demystifying technical expertise. Wherever controversial decisions arise nowadays there will undoubtedly be a few among the opposition ("local experts") who have a grasp of the technical issues and who also recognize when scientific and technological decisions are imbued with issues that affect social and political values. The fact that more and more people in industrialized societies, especially the United States, have acquired a university education in scientific and technical fields is tantamount to a decentralization of knowledge. This decentralization is in step with the trend toward decentralization in decision making and the growing interest in public participation in scientific and technical decisions.[20]

Volunteer experts have evidently been able to assimilate technical knowledge without adopting the values (and ideology) that are often associated with the scientific and technical professions.[21] They have sought to use democratic processes to express their values, opposing decisions that would impose judgments by so-called experts about acceptable levels of risk. It goes without saying that the role and impact of the bearers of technical knowledge is as fundamental to an understanding of technological development as are the artifacts of technology.

The social context in which experts are most prominent today is the introduction of new technologies, particularly large-scale technologies.[22] These technologies include power plants (both nuclear and coal-fired), missile systems such as the MX and cruise missiles, chemical complexes and refineries, airports, and the like. Even the highly sought after Superconducting Supercollider met with local opposition in the states considered as a site by the U.S. Department of Energy. Public opposition now almost always

accompanies introduction of large-scale technologies notwithstanding the continuing assurances of experts hired by industry
and government. This tendency has been observed in a number
of modern industrial societies with diverse institutional
arrangements.[23]

Standard operating procedure for introducing these technologies
has involved defining questions of social and political values as
technical problems to be dealt with by "the experts." This strategy
often worked in the past. Until the late 1960s the public seldom
questioned the value of technological progress. However, since
about 1970 public acceptance of this strategy has faded especially in
connection with the development of nuclear power. Attempts to
narrowly define issues as "technical" have created public hostility
toward experts and public apprehension about technology. The
impetus for the role of voluntary expert derives from the institutional
abuse of expertise.[24] Citizen groups have responded in controversies
by acquiring resources to enhance their capabilities in the political
struggle over information.[25]

The ideology of a countervailing expertise has been inspired to
some extent by the changing scale of social activities. Human
problems are no longer dealt with on a human scale. Distant, complex, and monolithic organizations of both government and business
increasingly seem to impose themselves over all facets of life and
every geographic area. These organizations have a technocratic bias,
acting as if only those with expertise can know what is good for the
state and its citizens.[26] The loss of control is soon apparent to those
who would seek to alter the decisions effected by these large-scale
organization. Dorothy Nelkin suggests that the question "who
controls?" is the main undercurrent in the rising ambivalence in the
public perception of science as a source of legitimacy. Controversies,
according to Nelkin, ". . . express both an ideological resistance to
the reductionism epitomized by science and a political resistance to
the powerlessness implied by the pervasive influence of expertise."[27]

The trend toward disagreements among scientists and technicians
and the concomitant rise of voluntary expertise has been viewed as
a threat to the community of scientists and technicians. This may be
part of the reason why very little attention has been paid to the study
of voluntary experts. The general orientation in controversies is to
find ways to get experts to agree (ergo, the science court), thereby
seeking to preserve the integrity of science as a source of consensus.
An equally defensible point of departure is that the rise of the
voluntary technical expert is democratizing science and technology
and that, as such, it is a healthy trend that bears encouragement. In

view of the ways science and technology have been institutionalized, it would seem that voluntary expertise is functional for democratic societies and accordingly, worthy of study and improvement.

The recognition that volunteer experts are a significant part of technical controversies has not been matched by an interest in exploring their contribution.[28] Whatever the reasons for this neglect there is good cause to study them. The character and focus of expertise has been changing as disputes shift from the national arena to local areas where "local experts" are increasingly called upon to defend social and political values. The impact of the volunteer expert appears to be much more significant at the local level. Although lacking in credentials and the respect of the scientific community, local experts have generally been legitimated in local controversies. Citizen groups have learned well that if expertise can serve as a weapon for social manipulation in the name of rationality, then it can be used to create conflict to serve their ends. The curiosity of social scientists should be aroused by the fact that citizen groups in such disputes very often succeed with the help of volunteer experts despite disproportionate resources available to project promoters.[29]

The Interplay of Facts and Values in Technical Debate

The main difficulty with the institutional abuse of expertise then is that questions of social and political values are defined as technical problems presumably best left to the experts. When volunteer experts are marshalled to dispute with the paid experts of government and industry, debate centers on who has the facts; but in reality, such controversies have little to do with disagreements over matters of fact. Value issues, even if not explicit in the controversy, are the essential basis for conflict.[30]

It is seldom possible to separate disagreements about facts from the value issues at stake.[31] If this separation is impossible, then it follows that experts cannot be relied upon to make decisions that the methods of science admittedly cannot resolve. The experts are no better off than the rest of us in rendering value judgments. When it comes to values, we are all experts.[32] Given that value issues are the prerogative of all concerned, separating facts from values is not a useful approach, either to managing conflict or to the study of expertise.

Technological controversies stem from factual uncertainties that allow for diverse and value-laden interpretations. Technical questions become controversial in view of the ambiguity between the technical and the normative.[33] The public is usually less interested in the technical facts under dispute than it is in the choice between different political and social values.[34] This is an especially important feature of environmental controversies.

Controversy highlights the need for a political solution that reflects community values. Complexity in technical debates is confusing to the public and foments additional controversy. The existence of controversy thus encourages the decisionmaker to revert to his or her original normative predisposition or simply to avoid the decision. During the course of a controversy acceptance of technical advice increasingly hinges on the extent to which it reinforces existing values. In this context opponents find a window of opportunity because it is often sufficient to raise questions that undermine technical expertise. The quality and quantity of the opposing evidence need not match that of the promoters.[35]

Strategies of Expertise in Technical Controversies

The dynamics of expertise in technical controversies is essentially a drama of conflicting facts and values. Expertise is used as a political resource for enhancing the power of decision makers. Regardless of what the experts may think of their role, this is the ultimate purpose that they serve. By means of experts and expertise political elites attempt to keep decisions that involve social values within their dominion, thereby excluding others with an interest in the values affected by such decisions. Citizens whose values are excluded search for a means to take part in the decisionmaking process. Because the conflict is joined in an arena of technical debate, opponents must provide counterexpertise to influence decisions. The resulting conflict between legitimate experts confuses the public and the decision makers. Where the path of technological development is not accompanied by consensus, decisions are withdrawn from the experts and made on the basis of politics. The political process has been so tempered that the mere existence of controversy is often sufficient to defeat a proposal or reverse a decision.

Many social observers have described what I term the strategies of expertise. This book is a systematic study of this social fact. To set the stage for this treatment, a variety of descriptions by scholars shows how widely recognized the strategies have become and gives force to the preceding generalizations. These excerpts from the literature appear below.

... two tendencies were at work in administrative environmental politics. On the one hand, there were those who sought to limit the arena of influence that affected decisions, to confine the actors to a smaller circle of scientific, technical, and professional people who could establish the terms of debate and agree on the numbers. On the other hand, there were those who sought to define such decisions as political choices of con- sequence and to widen the arena of influence about them. Administrative agencies sought valiantly to deflect such input, giving ground often in the formalities of environmental-impact statements, studies, and planning, but trying to present technical detail in such a way that the broader citizenry would have less influence than would the technical experts selected by administrators.[36]

The participation of technical experts has reinforced the existing tendency to define policy questions in technical terms. It has forced citizens whose interests are affected adversely by government policies to engage their own technical experts in order to challenge those policies.[37]

One of the most pervasive political impulses of the world of expertise is to drive the context of decision-making underground, beyond the purview of the general public, by imbedding it in enormous and complicated detail that makes it all but impossible to grab hold of. Those who wish a more open system of decision-making have a challenging task of bringing to the fore the choices involved in the way the problem is defined, the selection of variables to be measured, the weights to be assigned to the variables, and the treatment of the unmeasurable.[38]

The public scientist thus needs legitimacy from a neutral reputation but gains effectiveness from political action. The effective scientist in the area of public policy ... must know about the limits of scientific uncertainty, the demand for certainty by bureaucratic organizations, and the con- sequences of social conflict—and how to use those constraints. In sum, the scientist must know how to be a political actor.[39]

We ... observed a patterned recourse to expertise. Officials claimed to have knowledge and understanding unavailable to laymen. They said that citizens should entrust decisions to the experts. And citizens, unhappy or uncertain about these decisions, turned also to experts to attack official positions. This dependence on expertise involved a pattern that denied democratic procedures; experts would evaluate a problem and propose a solution that could not be debated in the public arena because technique was more important and more reliable than politics.[40]

Latent disagreements among experts concerning the legitimacy of competing sets of evaluative criteria constitute an all-important "hidden debate" that underlies and energizes the public number-slinging and name-calling that meanwhile masquerade as the essence of socio-technical controversy.[41]

... it appears fully justified that citizens' initatives and opposition groups insist on nominating their own experts, given the political power which accrues to those who can influence the rules of the game. The phenomenon of experts and counter-experts contradicting each other in public is therefore largely the consequence of the political forces which set the political stage for experts to carry out their function.[42]

Experts employed by proponents of development produced reports advocating growth and expansion, while experts engaged by public agencies or citizen groups opposing development warned of danger to people, property, and the environment. The conclusions of these studies usually coincided with the personal preferences of the experts, and they were not obviously inconsistent with the canons of scientific inquiry. However, they always satisfied paying customers, whether governments or private citizen groups.[43]

In cases of conflict all parties resorted to the essential weapon of technocracy, the expert. Experts proved, with rare exception, to be hired guns, irrespective of the identity of their employers or their area of expertise. Their results rarely deviated from the preferences of their clients. They added immeasurably to the cost of analysis and development, but their technical contributions are suspect, at best, given their ability to serve all sides of all disputes. Technique and expertise have altered the style of politics but have neither clarified issues nor contributed to compromise.[44]

As can be seen in these accounts, the strategies of expertise are not a radical change promoted by isolated groups that wish to stop technological progress. This behavior instead reflects a new populism created by changing values and the environmental consciousness of the last two decades. Neither are the strategies of expertise an unexpected development in view of the political use of experts by government and industry.

The ensuing study of wood energy development reveals yet another replication of this pattern, with some interesting details resulting from the factors studied and the methods used. Wood energy development is a type of siting controversy. Although concern about the quality of life in the community is the source of opposition in such controversies, the debate centers on technical issues. Opponents respond by seeking to manipulate knowledge, emphasizing areas of uncertainty that are open to conflicting scientific interpretation.[45] In this way citizen opposition groups aspire to justify their political and economic views. Political values and scientific facts consequently are difficult to distinguish in the conflict. The result, as Nelkin points out, is that "... in the various siting controversies no amount of data [can] resolve [the] differences. Each side [uses] technical information mainly to legitimate a position based on existing priorities."[46] Controversies over the siting of WFPPs would seem to be tailor-made for study of the strategies of expertise. The concepts of expertise presented in this chapter have laid the foundation for such a study.

2

Social Change as a Source for Controversy in Wood Energy Development

The Context of Change in Modern Industrial Society

A number of social changes have influenced the social reception of nonmetropolitan resource development including wood energy development. These changes are a reflection of the general context of technological change in the industrialized societies. Most important among these for nonmetropolitan areas are the population migration turnaround, the energy problem, the environmental movement, and the development and siting of large-scale facilities. Each is connected with the drive to develop wood for energy. This chapter establishes the basis for further analysis of large-scale wood energy development by summarizing the relevance of each of these important areas of social change.

The pace of change in rural communities in the United States has dramatically increased during the last two decades. Many factors are responsible for this ferment. Foremost is the turnaround in the rural-to-urban population migration, which has served not only to reverse the longstanding state of decline in many areas but to increase the pluralism of rural communities as well. The social and political life of these communities is being altered in the process. The population migration turnaround, as it is called, is a relatively recent social change. The migration from rural to urban areas during most of this century occurred for economic reasons. By contrast, the turnaround has in general been oriented to quality of life. The environmental values of newcomers are implied by their residential preference. Local politics have been stimulated by newcomers who aspire to influence decision making in their communities.[1] The turnaround in regions with scenic and recreational amenities functions as a vector for change, ushering in both the ideologies of energy development and local control and the strategies of expertise. It has been argued that changing values and rising living standards brought resources to communities that were utilized as political leverage.[2] Although presently in an attenuation phase, the turnaround will continue to affect nonmetropolitan areas for years to come.

Another major factor is the rush to develop energy resources generally found in rural areas. Industrialization of nonmetropolitan areas has been continuous in the United States. Nevertheless, the size and rate of recent development of energy resources represent a qualitative social, political, and economic discontinuity for rural areas.[3] The literatures of both the multidisciplinary field of social impact assessment and the sociological study of boom towns have become voluminous in step with the pace of development. Renewable energy resources as well as nonrenewable fossil fuels are high on the agenda of rural resource development. Qualitative community changes attending the influx of newcomers and the tendency for decisions to be made by outside interests have made development a complex problem for local governments. Many rural communities find themselves locked in a struggle to maintain control of their destiny.

The growing recognition of the environmental problem at the beginning of the 1970s was an irreversible change for industrial societies. There is no need to recapitulate the resulting history and impact of the environmental movement.[4] The persistence of public concern for the environment and a viable environmental movement are critical features of the current social context for large-scale development projects.

The major link between industrial society and the environment is in connection with energy production and consumption. The scale of energy use today is perhaps the most significant source of environmental impact created by human activities. Acid rain, the risk of reactor meltdown, and radioactive waste disposal are prominent examples of environmental issues in electric power production. Even proposals to build relatively small (by conventional power standards) WFPPs lead to protracted controversies. Although a renewable source of energy, development of wood power is apparently not immune to environmental concern when the scale of WFPPs portends significant environmental impacts.

Similar to the turnaround itself, development of energy resources has declined in recent years. Stable energy supplies and low prices for oil on the world market are responsible for this retrenchment. This should be viewed as a temporary situation. We know with certainty that fossil energy resources are being depleted and that domestic development will once again intensify. Efforts to develop wood-electric power persist in the United States largely because of tax incentives enacted when oil was more expensive and supplies were less reliable. The present study is relevant to this continuing development and should also contribute to understanding future controversies over energy development.

Environmental Awareness: The Michigan PBB Contamination

The growing public concern with the health of the environment can be illustrated with an example that had a direct effect on the effort to develop wood for electric power in Michigan. Hersey was one of the PBB contamination sites.[5] The political fallout from the handling of the PBB problem has significantly affected institutions in Michigan. The Michigan PBB contamination illustrates public recognition of the environmental problem and how it has changed public attitudes and environmental politics in the state.

The relationship between the PBB contamination and the response of the Hersey community to the proposed WFPP is not explored in detail. Rather, direct effects encountered by Hersey are mentioned based on media accounts and in post-hoc interviews with selected participants. The example thereby also serves to further describe the context of the Hersey controversy. Here follows a brief summary of the PBB incident and its legacy.

An important development in the public perception of environmental hazards in Michigan and, indeed, the United States, was the PBB contamination of 1973–78.[6] Every citizen of the state soon became aware of the potential danger of the chemicals being manufactured by modern industrial societies.

Two journalistic case studies have been written about the accidental mixing of PBB with cattle feed in Michigan which started in 1973.[7] The chemical worked its way through the food system with catastrophic effects on the farms and farm families who inadvertently used the feed. The resulting controversy peaked between 1975 and 1978 just prior to the Hersey WFPP proposal.[8]

Almost all Michigan residents now have measurable levels of PBB in their bodies. The long-term health effects have not been established. It is clear, however, that agriculture in Michigan has been transformed by the experience with the PBB contamination. Attitudes and perceptions about toxic chemicals and the need for institutional mechanisms to protect the environment lie at the core of this change.

Both case studies document the magnitude of the environmental impact and political haggling that resulted from the contamination.

Osceola County (especially the Hersey community) and adjacent Oscoda County contain clusters of farms affected by PBB. The map in the Egginton study names Hersey among several dozen heavily affected sites referred to in the study.[9] Chen's study also gives a high profile to Hersey.[10] It is a reasonable assumption that residents of Hersey were aware of the PBB contamination and its effects on local farms. They must also have been aware of the belated state attempts to shore up the problem. In any event, it became a major issue in the gubernatorial campaign of 1978.[11] The democratic challenger was given a good chance of winning because incumbent William Milliken had been so slow to give priority to the problem. Egginton notes that many outstate farmers were so incensed that they were prepared to vote democratic for the first time in their lives. Despite setbacks in the courts for victims of PBB, the issue persisted in the political arena. Governor Milliken was reelected but Republicans lost some of Michigan's rural vote and a few legislators who had supported the politics of the vested interests were ousted. The director of the Michigan Department of Agriculture resigned under pressure.[12]

Experts were solicited to decide if PBB was the cause of the problem; and when causation was established, what effects could be attributed to PBB. Experts from state government and the universities quickly lost credibility because they served and vigorously defended the regulatory institutions involved in various stages of the contamination. Egginton examined the role of experts, devoting several chapters to it, entitled "What the Experts Knew" and "Politicians versus Scientists." In an attempt to regain public confidence, Irving Selikoff, a medical researcher from New York and pioneer in the field of chemical contamination, was hired as an outside consultant to conduct an independent investigation.

The disposal of contaminated cattle carcasses became a hotly contested proposition in the northern Michigan community selected for the burial. Mio, a town in Oscoda County just to the north and east of Osceola County, bitterly opposed the disposal that eventually took place there. Oscoda residents fought the measure in court and when that failed, took such actions as scattering nails on the road to the clay-lined disposal pit. They displayed effigies of Governor Milliken, Howard Tanner (DNR Director), and Dale Ball (Agriculture Director) hung from a gallows at nearby road junction. This demonstration lasted for months.[13] Eventually the burial took place by court order. The state subsequently had the remaining barreled, frozen carcasses of contaminated cattle shipped to Nevada for disposal. Ironically, the 3500 or so animals buried at Mio contained an estimated two ounces of PBB among them. Earlier an estimated 269,000 pounds of the chemical had been dumped without protest into a Mio landfill between 1971 and mid-1973. The community of Mio had obviously become accutely aware of the chemical contamination pursuant to the PBB incident.

When the Hersey WFPP proposal was debated at a public meeting late in the controversy, a disturbance that almost halted the meeting followed the answer from plant promoters to question about whether or not PBB-tainted cattle would be burned in the facility.[14] Some of the residents of Hersey had also been radicalized by the handling of the PBB contamination.

Power Plant Siting and the NIMBY Syndrome

Siting of large-scale facilities has become the focal point of public opposition to technology. Such project are a convenient target for opposition because, as Casper and Wellstone note, "... the threat is tangible, the adversary is identifiable, and direct resistance is possible. Projects ... can become symbols of the larger threat to their existence that rural citizens are facing in this era."[15]

In general, the recipient community must bear the costs of a project that will largely benefit a different population. When called upon to accept this burden, communities increasingly refuse to sacrifice local interests. Many different kinds of facilities have generated this kind of response, including airports, high-voltage power lines, facilities for nuclear waste disposal, coal-fired power plants, and most expecially nuclear power plants. Defense facilities such as the proposed MX missile deployment in rural areas of the western United States and the ELF submarine communication grid planned for upper Michigan and Wisconsin have been opposed on similar grounds.[16] More recently public protest has occurred in a number of states being considered for the Superconducting Super-collider. The rejection of radioactive waste disposal sites is almost universal. It appears at this juncture that many states will seek to withdraw from compacts with other states in order to avoid the issue of site selection for low-level radioactive waste disposal.

Disputes over the siting of power plants are most representative of this type of public response. However, public reception depends in large measure on the kind of technology. Opposition had less frequently attended the siting of fossil fuel power plants, perhaps due to long experience and familiarity with them. Public reception of nuclear power plants provides a sharp contrast. Controversy began soon after the start of domestic nuclear power development in the United States, building to a fever pitch after the infamous accident at Three Mile Island in March 1979. The recent accident at Chernobyl, USSR may have sealed the fate of nuclear power development. The obvious policy relevance of power plant development commends social science research on the public reception accorded different kinds of power plants to make findings more generalizable.[17]

Perhaps the most feared response a developer can encounter is "not in my back yard."[18] The acronym NIMBY has been combined with the notion of syndrome to give opposition to facility siting a convenient label. The idea of NIMBY syndrome evolved from the issue of radioactive waste disposal siting.[19] A substantial literature has been generated by this social response.[20] The NIMBY syndrome derives from the fact that the perceived risks of environmental and technological hazard have been brought close to home, directly impinging on individuals, families, and communities. NIMBY has led to the frustration of siting efforts throughout the United States and other industrial countries.[21] This political response to hazardous large-scale projects has sharply curtailed the activities of very powerful public and private interests.

The NIMBY syndrome is not as simple and straightforward as the term implies and stereotyping this response would be a mistake. As Wolf states, "the superficial attraction of the term masks a deep structure of conflict in personal and public choice."[22] Citizens do not take this position simply because they are against technology or the centralized bureaucracies of industry and government. Rather, they are concerned about loss of control, loss of community, and the inequities created by such developments. Preserving local values is the uppermost concern. The response is, therefore, not simply a reaction *against* the technology or its hazards. Rather, it is *for* maintenance of community values. What appears to be a reflexive, negative response must also be construed as a thoughtful affirmation. This affirmation takes the form of defending local autonomy, a topic to be taken up later in the chapter.

Rural areas occupy a special niche with respect to the NIMBY phenomenon. The pressure to site noxious facilities in rural areas has increased in proportion to public aversion to these facilities in urban areas. This historic change in described by Hays:

> Industrial entrepreneurs found it increasingly undesirable to locate new facilities in or near metropolitan areas because land was limited and public opinion adverse. They chose to shift sites to the countryside. Often

these new locations were thought to be in "remote" areas and hence not likely to generate public opposition. But to significant numbers of people few areas were remote. Those who had long called these places home now developed a new awareness of their capacity to resist, and they were aided by more recently arrived neighbors.[23]

The Population Migration Turnaround

The general phenomenon of net migration into nonmetropolitan areas that formerly lost population through out-migration is relatively recent. It has come as something of a surprise to interested researchers as well as to the communities themselves. The nature of the turnaround is considered here along with its effects in recipient communities.

The 1970s witnessed an unanticipated increase in the non-metropolitan U.S. population. Many of the growing counties had completely rural populations and were not adjacent to Standard Metropolitan Statistical Areas. Most of this growth was the result of net in-migration. The Upper Great Lakes Region and especially northern Michigan contain many such counties. For rural areas that previously lost population, the change has been dramatic, and the effects remain to be thoroughly studied.[24] In a recent account of the population deconcentration in the United States, Herbers summarizes the situation in this manner:

> ... this scattering of much of the population, as well as changed economic and social conditions, is having an important impact on the way we view our nation and the world beyond, on the kind of governments we elect, on the use of our resources, on the environment, on the development of transportation systems, and on the way we spend our time and money. Because development or expansion of communities in the outer reaches is a more radical step than the opening of the suburbs a generation ago, its potential for further change may also be greater.[25]

Growth in nonmetropolitan counties has been characterized by type of economic specialization. Energy extraction, recreation, manufacturing, retirement, and government-related activities account for most cases of nonmetropolitan growth. This finding suggests that type of growth will affect the type of migrant attracted to the area and the pattern of distribution of population and activities that accompany growth. A considerable literature reinforces the importance of these activities. Studies that point out the role of various rural-based amenities in the decisions of migrants to settle in these areas are especially relevant. A survey of residents in the growing counties of the Upper Great Lakes region found that the vast majority (75 percent) of in-migrants from metropolitan areas (comprising 85 percent of all in-migrants in the sample) gave noneconomic reasons for their move. The most important reasons were those reflecting pro-rural/anti-urban sentiments and preference for rural areas as locales for recreation and retirement. Others have also found such quality-of-life variables to be highly important in the decision to migrate.[26] The apparent desire for rural-based amenities is suggested by the kinds of areas in which migrants choose to reside: "... they are settling in areas rich in natural attractiveness, with large tracts of woodlands, hundreds of miles of shoreline and a clean spacious environment." Less than 40 percent of these migrants live within the limits of a city.[27]

Retirement and recreation are important activities connected with the increase in migration based upon natural amenities. Studies of rural areas have demonstrated that both contribute to population growth. A few of these studies have attempted to examine the relationship between place characteristics and the growth of areas due to recreation and retirement. For example, a study of second home ownership in the United States documents considerable increases in recreational land subdivisions, recreational lots, and second homes.[28]

The importance of vacation homes as a factor contributing to population growth is evident in the high rate at which second homes are converted to permanent dwellings. This trend is especially dramatic in recreational areas undergoing population growth. In some developments such housing is marketed as both permanent and vacation housing. Besides contributing to growth, vacationers have also served as allies in protecting the natural environment.[29]

The pursuit of environmental values is thus a significant feature of the turnaround. The environmental values of newcomers are implied by their residential preferences. The recent migration to nonmetropolitan areas largely stems from concerns about quality of life.[30] Newcomers and oldtimers alike value the amenities of the rural environment.

Facilities that threaten cherished values are subject to intense local conflict. To cite a relevant example, turnaround migrants have been instrumental in halting plans to site power plants in nonmetropolitan communities.[31] The task of energy facility siting has, therefore, become more difficult in the wake of the population migration turnaround. The turnaround has added to the complexity of rural social change. Traditional, homogeneous rural communities are in the midst of changes that are blurring the distinction between urban and rural. Hays put the change this way:

> Cooperation occurred between older rural residents who now spoke in terms of "quality of life" and more typical environmentalists of urban origins. Such cooperation, however tentative, was more than fleeting and isolated.[32]

Effects of the turnaround on receiving communities are both quantitative and qualitative. Quantitative effects are the sort generally associated with population increase, for example, the need for more and better services of all kinds. Qualitative effects have to do with more subtle changes. The quantitative account cannot provide a complete picture; studies have yet to focus on concomitant changing values and their influence in community life. Certainly, the turnaround brings *more* people to rural areas, but it is the fact that they are qualitatively *different* people that leads to conflict with the project developers.[33]

An example serves to illustrate the role of newcomers:

> The vitality of the anti-MX group came largely from its diversity.... [Leaders] were better educated and more likely to hold professional positions. Several of these leaders were also fairly recent inmigrants to the area. They could be characterized, on the surface, as exhibiting what has come to called the "last settler syndrome." That is, they have lived in urban America, haven't liked it and so have moved to a rural community to find a lifestyle that is more pleasing to them. Having found this lifestyle, they want to close the door behind them so that it is not disrupted by others who will follow. They joined forces with ordinarily pro-growth locals to oppose the threat of the MX to local values.[34]

Rural values increasingly coexist with the urbanite values of inmigrants in remote nonmetropolitan areas as a result of the turnaround.[35] Whether or not the analysis of these values is the key to a new concept of rurality, the turnaround presents a laboratory for studying the function of values in social change.

Not all rural areas have been significantly affected by the turnaround. Many, in particular the farm belt and areas that are economically depressed due to closure of manufacturing plants, mines, oil fields, etc., lost population in the 1980s. The fact that certain natural resource areas have turned around is central to this study. Such areas contain the forests being considered for development to fuel WFPPs. Yet even in some of these locations the turnaround has attenuated recently. As a result, the same scholars who were skeptical about the existence of the turnaround in the first place now seem anxious to conclude it is over.[36]

There are three difficulties with this point of view. First, effects of the turnaround should continue to be felt in recipient communities for years to come even if the migration ceases.[37] Second, changes in census definitions and a misinterpretation of the data appear to be responsible for the idea that the turnaround was transient. To quote Herber's recent study of population deconcentration in the United States:

> within five years after the 1980 census, surveys of migrations and population estimates showed metropolitan areas once again were growing faster than nonmetropolitan areas. Some demographers were quick to say the "rural rennaissance" of the 1970s was over and people once again were moving to the cities as they had done during most of the nation's history. This, in my view, constitutes a misreading and misunderstanding of the facts and figures, as the Census Bureau findings say little or nothing about the movement of people and commerce to the countryside. Metropolitan areas now have such extensive boundaries that they cover about one-fifth of the land area of the United States, and most of that land area is rural or semirural, not urban. And it is rural land within metropolitan areas that is now having the greatest population growth. After the 1980 census the federal government reclassified so many small communities from nonmetropolitan to metropolitan—even when they had few if any metropolitan characteristics—that the shift of the 1970s, which was a shift from central cities to far outlying areas, cannot fairly be compared to what is happening in the 1980's . . . . *All of the population estimates now show that for the first half of this decade small metropolitan areas that most people have never heard of and the outlying counties on the fringes of large metropolitan areas are having much faster growth than either the old cities or the established suburbs.* [Emphasis mine.][38]

Third and finally, migration is a *net* assessment of population change. Even in instances when rural areas are depopulating due to out-migration, there may be a migration stream in the opposite direction that brings urban defectors to the country. A study of the MX missile controversy in Utah and Nevada demonstrates that even in regions that are losing population newcomers may be present to lead opposition to controversial developments.[39]

The turnaround has been somewhat volatile, but this is all the more reason to look into its dynamics and social effects. The evidence presented in this study suggests that the turnaround is the major avenue for social change in nonmetropolitan communities, particularly in regions with extensive forest resources for wood energy development.

Ideologies of Local Development and Autonomy

The tendency for decisions affecting rural communities to be made by outside interests is increasingly apparent in local affairs. Growing dependence on outside interests and the corresponding loss of local autonomy is, perhaps, the most significant ideological factor in siting disputes. Project promoters appear to embody such interests. The forces generating this concern are both public and private. Federal and state governments have increased their involvement in local affairs largely because many problems supersede local boundaries, or solutions do, or both. Environmental problems typically require attention by higher levels of government. Air and water pollution, for example, do not honor administrative boundaries. Moreover, environmental problems are often exported to the countryside because of public resistance to their resolution in urban communities. Hays points out that ". . . more highly developed areas [prefer] not to absorb the cost of environmental impact, [and] instead [seek] to export that cost elsewhere."[40]

The ideology of local control in siting disputes rests on the paramount concern that the local community may be called upon to bear the social costs of a project imposed by groups outside the community.[41] Asserting the privilege of deciding what happens in one's local community is an ideological position identified as the NIMBY syndrome just discussed. As such it is a plan of action for project opponents. Very often controversies get down to the issue of local autonomy especially if the pleas of citizens groups are ignored by the agencies of state and federal government.

By contrast, promoters of large-scale technological developments usually take a financial interest in a project. Their orientation to siting disputes rests on the ideology of development. Developers proceed with a view of economic rationality presumed to benefit society as a whole. Usually their strongest argument in favor of a facility is the benefit of the project to the local economy. Very often they acquire the support and assistance of the local business community whose economic orientation is toward the large urban centers of capital and organizational control. In response to pressure developers may get into discussions of environmental tradeoffs versus economic benefits. Yet, project promoters generally try to avoid the direct approach, instead employing expertise to mask political choices. By this means they have sought to control decisions that are crucial for local communities.[42]

The general concern for environmental quality and its formalization in the two paths energy policy debate is reflected in the myriad of local disputes over large-scale energy developments. The debate about which energy path we should follow implies an important connection between local disputes and more cosmopolitan concerns.[43] The distinction between local and cosmopolitan issues is also useful for circumscribing the conflicting ideologies of local autonomy and development.

Protest against technological developments usually arises in urban communities where cosmopolitan concerns motivate the opposition. The importation of cosmopolitan perspectives into nonmetropolitan communities is argued here to be an essential ingredient of WFPP controversies. The mechanism for this change is the population migration turnaround. While newcomers who oppose developments may proceed from a cosmopolitan perspective, they strategically use the ideology of local control to gain legitimacy and support from the local community.

Technology is, in general, disruptive of older norms. Protest groups fight to preserve values lost in the course of technological progress.[44] One's response to technological change depends on his or her perspective. Long-time residents and rural in-migrants greatly esteem the physical and social amenities of the rural lifestyle. This value has both cosmopolitan and local dimensions. Environmental values are part and parcel of this attitude toward local amenities. Newcomers and oldtimers are generally found to be in agreement on this subject.[45]

Residents of nonmetropolitan areas often are acutely aware of their situation vis-à-vis the centers of power and capital. The following statement from a participant in a siting dispute articulately summarizes the local perspective in such controversies:

> Most of the people live in urban areas. Most of the decisions are made with an urban bias and very much to the detriment of the countryside. Anything that can be ripped off the countryside and sold in the city, is. They lack a certain element in their consciousness that we need. We need an urban-rural balance. They've gotten into a very exploitive-type posture. It's hypocrisy to think that we should now build power plants in the countryside because the urban areas . . . need clean energy. It's an exploitive attitude . . . . We ought to have the right to our way of life too.[46]

Local perspectives on environmental problems are greatly influenced by belief about cosmopolitan issues. Opposition leaders in particular are impelled by concerns with larger national problems. In energy facility siting controversies, decentralization of technology and its sociopolitical accoutrements is an issue used even by project promoters to make their case.[47]

In such technical controversies project promoters ignore local interests and, based upon a developmental ideology, seek to define decisions as technical and to impose these decisions on the local community. One group of scholars provide an especially cogent discussion of the source of this tendency. They are well worth quoting at length here:

> When acting with statutory or corporate authority, [public and corporate] officials characteristically assume a role of rectitude. Thus they tend to view the interposition of so-called public interest groups (e.g., on behalf of environmental protection) as obstructionist meddling. The acrimony characterizing many environmental disputes reflects moral indignation felt on both sides of the controversy. The environmentally concerned citizen sees the public or business official as sacrificing or betraying the interest of society at large for some special economic or political purpose that is substantively or morally wrong whatever its legality. The officials see the protesters as self-appointed troublemakers interfering with the orderly, lawful, and efficient processes of government and business. By what right, they ask, does this self-interested minority burden and delay the conduct of public business for which society, in its wisdom, made them and not the protesters responsible?[48]

Apparent in this statement is the strategy of citizen opposition groups. Motivated by cosmopolitan concerns, they mobilize local resources to defeat projects they regard as serving special interests.

The Relevance of Social Change to Strategies of Expertise

The use of experts and expertise in environmental politics is relatively new. It emerged into a world of social turmoil characterized by changing values, perceptions, and attitudes. Longstanding social patterns have been abruptly reversed. The net population increase in many nonmetropolitan areas due to in-migration, for example, caught many scholars by surprise. Such changes have been uneven from one geographic region to the next, but nearly every corner of American society has been affected. Understanding the politics of expertise requires some attention to these changes.

Neither is the response of the public to technological change and the behavior of the public in technological controversies divorced from these changes. To understand the dynamics of controversy and the role of experts and expertise, it is important to consider the social changes that influence them. This is comparable to exploring a digital circuit board with a logic probe. The flow of current is not obvious just from observing the connections. By the same token, probing the intricacies of social change requires a long period of study, careful observation, and a good head for discerning relationships where they actually exist. The researcher must keep in constant touch with the work of others who study the subject. The process often benefits from time and further reflection. In the meantime new data on social change inevitably crop up.

This study of wood energy development covers a period of about ten years. The original case study and the cases examined since then reinforce the importance of social change in wood energy development controversies. Siting disputes remain the foremost source of public outcry over the environmental effects of technological development. The population migration turnaround has attenuated somewhat, but the new class of in-migrants continues to affect local politics in many nonmetropolitan communities. Their skills, attitudes, and action have proven useful time and again in dealing with large-scale projects that threaten community values.

The context of change has been summarized in this chapter together with an extended discussion of important social changes. The social changes described are critical to the ensuing trans-disciplinary account of large-scale wood energy development controversy.

Part II
Wood Energy Controversies

3
Wood Energy Development:
Concepts and Issues

Wood as Energy

Wood is a form of solar energy. Forests are an essential means of collecting and storing the energy of the sun. Wood is also humanity's oldest energy source. This ancient fuel is today described by the term "biomass"—any energy source that relies on living (organic) materials.[1] Because wood is derived from living systems it is regarded as a renewable resource. In principle wood that is harvested for energy or materials can be renewed by planting and managing forests. Foresters refer to this as the sustainable yield. For wood to remain renewable no more of the stock can be taken in a year than is replaced by new growth.

To understand wood energy development it is important to properly view the nature of forests. Land is far from being space where relentless improvements in technology will yield resources and energy without limit. The limits of land to produce wood are circumscribed by biological and physical laws. Forestland has economic value for the same reason as a fisherman's net—because it captures energy.[2] Some of this energy has been stored during prehistoric times to become the essentially nonrenewable fuels of the industrial revolution.[3] We are inexorably running out of non-renewable fossil fuels. Recognition that fossil fuels are being depleted is the basis for the current interest in alternative energy sources including wood.

Wood was the primary fuel in the United States until the mid-1800s. It was second to coal for the rest of the century. The peak of wood fuel consumption came in about 1900. At that time over three hundred million cords of wood were burned annually in the United States.[4]

The use of wood for the production of steam and/or the generation of electric power is not new in the United States. The

Wood is a form of biomass energy. Trees convert solar energy into biochemical energy by means of photosynthesis. Direct burning of wood is the most common method for utilizing this energy source. *Courtesy of the Minnesota Department of Public Services, Energy Division.*

practice was common early in the century when wood represented a much larger proportion of energy consumption. Household use of wood was also commonplace. In 1900 40 percent of all roundwood cut was used for heating and cooking. By 1970 less than 2 percent of all U. S. households used wood as a primary source of heat.[5]

The modern use of wood biomass for power production has largely occurred in the forest products industry of the western United States to provide industrial process steam. The municipal utility of one of the towns in the region, Eugene, Oregon, has made use of surplus forest residues to generate electricity and steam since 1941.[6] As of 1979 the Eugene Water and Electric Board was able to sell four hundred fifty thousand 1b./hr. of steam and generate up to 34-MW of electricity.[7]

Wood is an ancient source of fuel that experienced a resurgence of interest in the United States during the energy crises of the 1970s. *Courtesy of the Minnesota Department of Public Services, Energy Division.*

The arrival of the energy crisis in the 1970s directed attention to alternative sources of energy including wood. The greatest increase in use of wood for fuel has been in the residential sector for space-heating. By 1982 more than two million U.S. residences used wood as their primary source of fuel.[8] The national increase in residential wood burning has been put at 300 percent during the period 1973–1980.[9] The use of wood by the forest products industry has also accelerated. This industry is by far the most significant consumer of wood energy in the United States. Use of forests for energy is a practice that was nearly extinct in the 1960s, but today forests are once again providing wood for energy.[10]

Interest in use of wood as a fuel to generate electricity also

Forest products industries in the United States have long used wood waste for power. More recently residues from these industries have been utilized to produce electric power. Pictured is the Pacific Oroville power plant, located in Oroville, California, which relies primarily on wood products industry residues for fuel. It opened in September 1985 and is rated at 18.7 MW. *Courtesy of Pacific Energy.*

increased in the wake of the energy crisis. The Northeast United States has been on the vanguard of wood-electric power development largely because of its dependence on imported oil. The hamlet of Dixville Notch, New Hampshire became one of the first communities in the country to be completely lighted and partially heated with wood by utilizing wood chips from local logging and pulp mill operations.[11] Burlington, Vermont has been on the forefront of this development. It retrofitted a 10-MW coal-fired power plant in 1977 to burn wood for fuel. A favorable experience with this facility soon led to consideration of a 50-MW WFPP. After a brief period of indecision the plant was built, becoming operational in 1984.[12]

At about the same time as the 50-MW facility was proposed for Vermont, the state of Michigan promoted a demonstrational power plant to begin large-scale development of the state's forest for energy production. Hersey was ultimately chosen as the site of the plant. The ensuing controversy resulted in defeat of the project and set the stage for persistent public concern about the environmental impact of wood-electric power development in Michigan. The study of this controversy begins in the next chapter.

Wood as an Alternative Source of Energy

The essential long-term problem for energy policy is to divine a path from nonrenewable fossil energy resources to renewable alternatives. This is the sum and substance of the energy problem. Wood energy is one such alternative, but unlike fossil fuels it is not conveniently concentrated in one place. The fact that wood is geographically dispersed on land that also provides scenic and recreational amenities confounds efforts to develop wood for energy. Rural landscapes are the object of environmental values, particularly as the encroachment of industrial society threatens to degrade them.[13]

Consider as well that optimal location of a WFPP to minimize transport costs dictates that it be sited near the area to be harvested. For obvious reasons this location is likely to be rural and remote. Wood energy development is certain to be more problematical for rural communities.

Development of wood energy resources has been heralded at a time of rising public disenchantment with large-scale energy producing technologies. Concerns about risks and environmental hazards have virtually halted nuclear power development since the 1979 accident at Three Mile Island and the subsequent meltdown of a reactor at Chernobyl, USSR. Air pollution from coal-fired electric power plants remains a significant source of public concern. Although wood energy presents risks and hazards of a lower order of magnitude, centralized schemes for wood-electric power production are also being carefully scrutinized for their prospective impacts.

The policy debate in energy development has not centered on the choice between renewable and nonrenewable. While it is true that in general renewable energies and conservation have been pitted against conventional and nuclear power plants, the debate hinges on a more generic issue. The choice has been posed between large-scale, centralized ("hard") path technologies on the one hand, and small-scale, decentralized ("soft") path technologies on the other. By this reckoning it is the manner in which development is organized and the scale with which it is implemented that counts. The fact that a resource is renewable does not assure its development will be unopposed. Large-scale schemes for developing alternative energy resources have been criticized right from the start. For example, very large windmills have received an unenthusiastic reception. Some attempts to install large wind generators have failed.[14] Satellites for converting solar energy and beaming microwaves to earth is another example of a large-scale plan that has been opposed.[15]

The hard path/soft path debate rests on ideological precepts and ultimately on values. The community effects of scale and centralization on environmental values inspired the debate. As this study of wood energy cases unfolds the importance of these sociological factors and the manner in which they affect controversy will be described in detail. The final section of this chapter will return to the ideologies of energy development and describe the two paths ideological debate in finer detail.

Regardless of the scale of development, wood appears to be environmentally benign when compared to nuclear or even coal-fired electric power plants. Indeed that the wood-electric alternative was proposed in the first place implies the desire to increase our energy supply without added threat to the environment. There is, however, a quantitative difference in viable scale between wood and conventional power plants. Fuels for conventional power plants are concentrated and thus can be economically transported to a central location. Economies of scale exist under these circumstances, making power production cheaper for larger power plants. Wood is both a widely dispersed and less concentrated energy source. It must be harvested from a relatively large geographic area. As a consequence the transportation costs for wood are high, restricting

Wood chips are a versatile form of biomass energy suitable for burning in power plants. *Courtesy of the Minnesota Department of Public Services, Energy Division.*

production facilities to densely forested regions.[16] As a result of such economic factors, the feasible scale of WFPPs is of a much lower order of magnitude than conventional and nuclear power plants.

Economies of scale may exist for WFPPs up to 50 MW, but economic constraints on wood harvesting and transport quickly limit feasibility of larger plants. Coal and nuclear power plants are usually at least ten times larger. Their economies of scale are not dependent on available harvest and transport distance for a biological resource containing less concentrated energy.

Experience with the economies of scale has made it difficult for developers to accept the feasible scale of wood energy development.

A wood chip harvesting system. A fellerbuncher is shown harvesting a tree. A grappleskidder drags trees to the chipper, which in turn processes them into wood chips. These are blown into a van for transport. *Courtesy of the U.S. Forest Service.*

They tend to think in terms of the largest and therefore the most economical power plant, but this reasoning does not work with WFPPs. The resource is, by nature, different. Plants that are 25 MW or larger in size quickly run up against the problem of energetic and economic costs of transporting wood from a distance.

Combustion of wood, like other fuels, is relatively inefficient in converting the energy in the fuel to electric power. Direct burning of wood for electricity generation is 25 percent efficient a best, a figure substantially lower than coal due to the moisture content of wood.[17] Studies have demonstrated that use of wood for electric power production by utilities is not likely to compete favorably with coal, with the caveat that local conditions and requirements may favor wood for this purpose. The process of cogeneration—which increases the efficiency of the system by using waste heat—makes wood a more competitive producer of electricity.[18]

There are qualitative differences as well between wood and other forms of energy. The radiation hazard of nuclear power plants and the risks of system failure set it apart from power plants that rely on combustion processes. The virulent opposition to nuclear power stems from recognition of this difference. Concern about acid rain had made coal a less attractive source of energy. The most salient qualitative difference is that wood, unlike nuclear power and coal, is a renewable energy source.

Production of electricity creates environmental impacts no matter what the fuel. The mining of coal and uranium are a significant source of environmental degradation. Similarly, among other impacts, wood-electric power generation involves large-scale removal of the standing stock of nutrients in forest ecosystems. The effects of harvesting wood for energy is undoubtedly its most important drawback. In the controversies examined in this book, the impact of harvesting the forests has been a fundamental concern among citizen opposition groups. Wood harvesting can have beneficial impacts as well, which explains why public officials responsible for management of natural resources have often promoted wood-electric power.

One of the motives for development of wood is that it is low in sulfur. Sulfur (from combustion of coal) is a major contributor to acid rain. Because of its high moisture content, small amount of incorporated nitrogen, and low flame temperature, wood burning emits relatively low levels of nitrous oxide.[19]

Despite these advantages, wood is far from being environmentally benign. A number of pollutants in wood smoke have been established as hazardous to human health. Concern has focused on residential burning of wood but large-scale wood-burning facilities can also produce these pollutants. Up to fourteen carcinogenic (cancer promoting) compounds, six cilia tonic and mucus coagulating agents, and four cocarcinogenic agents have been identified in wood smoke.[20] The major pollutant from wood combustion is particulate emissions, requiring control in all larger units.[21] The particulate emissions from wood in the form of small bits of ash and soot exceed those for most forms of coal and are twenty-five times that for oil on a per-unit-of-heat-produced basis.[22]

Some aspects of the economic and energetic efficiency have been touched on in this discussion along with an array of environmental concerns about wood harvesting and wood-fired power plants. If environmental values have served to inspire opposition to wood-electric power, then economic values have been at the heart of the desire by promoters to develop it. Accordingly, the last item on the agenda is a brief discussion of economic issues that transcend both economic feasibility of the WFPP itself and the cost comparison of wood versus conventional fuels.

Wood power has generally been promoted in terms of the number of jobs it will produce in the area where the plant is to be sited. Because densely forested areas tend to have high rates of unemployment, the proposal becomes a dilemma for host communities. Residents are forced to balance environmental costs against economic benefits. A growing concern and perhaps the most important economic issue is competing uses for wood. Burning wood for energy is a relatively inefficient, low grade use of the resource. Residential wood burning is much more energetically efficient. At least some of the wood harvested would be more suitable for other commercial purposes. Developments in biotechnology are likely to increase the market for what is now considered waste wood. In the meantime, wood products manufacturing generates more employment per ton of wood than does wood energy production. Consequently, development of wood for electric power may have the unwanted by-product of reducing potential employment opportunities in the region.[23]

Prospective Issues in Wood Energy Development

A digression into issues in wood energy development will prove useful further along. Many have already been alluded to in the foregoing discussion. The following systematic listing of development issues points up many of the facts and values that surfaced in the wood energy controversies studied.[24] Not all of the issues listed arose in the controversies examined in this book. The purpose here is to set out the domain of technical issues from which controversy can develop. Familiarity with these issues will better enable the reader to distinguish and appreciate the technical and value arguments made in the wood energy development controversies studied. Four categories of issues are considered—economic issues, forest resources issues, potential impacts of large-scale tree harvesting, and environmental issues.

The array of issues in wood energy development provides many opportunities for technical experts to make value judgments. To the extent that these value judgments are useful to political elites, the expert is something more than a scientist or technician standing apart from the political process. We shall see if experts do, in fact, make these judgments under the guise of scientific propriety, and what effect this behavior has in technical controversies.

ECONOMIC ISSUES RELATED TO WOOD-ELECTRIC POWER DEVELOPMENT

Economic arguments are most often used in promoting large-scale development projects. Such projects have multiple economic advantages especially for nonmetropolitan areas that are economically depressed. Some of these *advantages* are:

- Jobs added directly to local economy
- Service jobs added to local economy
- Addition to tax base
- Multiplier effect of income provided by plant
- May generate regional employment
- Reduces amount of money lost from the economy in payment for imported petroleum products

There can be economic *disadvantages* as well, including:

- May reduce potential employment is some regions, e.g., due to effect on tourism
- High accident and death rates in logging
- Wood harvested for energy useful for other purposes
- Threatens the supply of lumber, pulp-wood, pulp, and other wood products
- Jobs lost by local, independent loggers
- Increased price of commerical wood
- Industries may have to reduce their pollution due to federal regulations so that pollution in the region does not exceed air quality standards
- Pollution aspect may discourage new industry
- Pollution may have an adverse impact on tourism or other local businesses, thereby reducing income in the area

Environmental factors are intimately tied to economic factors, a notion implied by the last three items in this list. It is important to note that economic arguments cut both ways. A power plant may in some respects be a boon for the local economy and a bane in others. For example, if the local economy is dependent upon tourism, the net effect of a power plant may be adverse for the community. Economic gains from the power plant may be more than offset by losses if the environment for recreational activities is degraded. Air pollution may also discourage new industry by reducing the regulatory carry capacity of the region. Once the regulatory ceiling on emissions for the region is approached, new industry may be prohibited from siting there or face burdensome costs to minimize air pollution.

Selective forest thinning is useful for forest management as well as for harvesting wood to produce fuel. *Courtesy of Pacific Energy.*

FOREST RESOURCE ISSUES IN WOOD-ELECTRIC POWER DEVELOPMENT

Public concern with forest management is already intense because of the recreational uses for forests. Changing values have established new criteria for managing forest resources. This issue soon rises to the top of the agenda when development of forests for electric power is proposed. As it happens, harvesting of wood for electric power production can have beneficial effects on forests. Prospective *advantages* of large-scale harvesting include:

• May promote more intensive management and better environmental practices in the forest products industry

A wood-chipping machine stands idle in front of a pile of wood chips. *Courtesy of Northern Logger and Timber Processor.*

- Conservation-minded cutting may help promote forest growth and wildlife diversity
- Removal of low quality species in woodlot may serve to improve the productivity of the stand
- Decrease in forest fire hazard through removal of debris

As you might imagine there is much longer list of potential negative effects of harvesting wood for energy. *Disadvantages* may include:

- Destruction of forests
- Deforestation
- Depletion of growing stocks
- Failure to generate new growth
- Reduced forest productivity
- Ecological change in forest character
- Damage to marginal woodlands
- Most affected sites will have fewer years to recover before they are again logged
- Erosion and accelerated leaching
- Loss of organic matter to soil through removal of biomass
- Damage or loss of ecosystems
- Burning is one of the least efficient uses of wood fiber

Wood chipping is a highly mechanized operation generally done on the site where the trees are harvested. *Courtesy of the Minnesota Department of Public Services, Energy Division.*

A typical woodchipping operation. The grappleskidder on the left drags felled trees to the wood chipper for processing. Wood chips are blown into the van for transport to the power plant. *Courtesy of Northern Logger and Timber Processor.*

POTENTIAL IMPACTS OF LARGE-SCALE TREE HARVESTING

Of greatest concern to nonmetropolitan communities rich in forest resources are effects on the forest itself. Virtually no aspect of the forest ecosystem may remain unaffected by the radical changes produced by large-scale harvesting. Some of the concerns in this area include:

- Soil compaction, disturbance, and erosion
- Soil moisture and structure changes, especially in thin soil areas of the boreal forest zone
- Increased organic matter decomposition and loss
- Loss of N_2 fixing bacteria and other nutrients
- Altered forest succession
- Wildlife species changes
- Monoculture problems with disease and pests
- Water pollution from pesticides and chemical fertilizers
- No slash left to stabilize slopes
- Increases in forest land area must come from current agricultural land

ENVIRONMENTAL ISSUES RELATED TO WOOD-ELECTRIC POWER DEVELOPMENT

There are numerous environmental advantages in using wood as a fuel. The popularity of wood fuel for residential heating and the interest in using it for power production both derive from these multiple advantages. The *advantages* of wood fuel include:

- Renewable resource
- Little or no sulfur or radioactivity
- Low levels of nitrous oxide emissions
- Does not exacerbate acid rain problem
- Use of waste (in forest industry)

The interest in wood as an alternative source of energy has been so great that its drawbacks have either been ignored or neglected. Many of the problems with woodburning have been discovered only recently. The general environmental *disadvantages* of wood fuel include:

- Air pollution
- High ash content
- Very high levels of polycyclic organic matter (POM)
- Health hazard

A whole-tree chipping operation like the one pictured above unavoidably removes nutrients from forest ecosystems. *Courtesy of Northern Logger and Timber Processor.*

- Emissions contain various toxic, irritating, and carcinogenic agents
- Contributes to acute and chronic health effects of air pollution, including chronic and acute bronchitis, common cold, pneumonia, emphysema, asthma, cancer
- Both soot and smoke are carcinogenic to the skin and lungs
- Pollutants can cause poor visibility
- Water pollution from soil and nutrient runoff in forests harvested for energy
- Soil depletion in forests harvested for energy
- Water pollution created by runoff from stockpiles of woodchips, ash, and other conversion and combustion by-products
- Destruction of wildlife habitat
- Visual change in forest character

Some of the issues defy attempts to categorize them. Visual change in forest character, for example, is an aesthetic issue but it is also an issue for forest management and the local economy in tourist-dependent regions. Suffice it to say that linkages between the issues enumerated are great. After all, the forest is an ecosystem. The web of interrelationships dictates that any one change can have diverse effects in the ecosystem.

By the same token, the complexity and interrelatedness of the economy portends diverse effects among these factors. For example, rising fossil fuel prices will probably lead to use of forest products as a replacement for more energy intensive materials.[25] The complexity and attendant uncertainty of technical choices can serve both to stimulate controversy and to provide strategic information for bringing controversy to closure. The next section looks more specifically at the ideological underpinnings of wood energy controversies.

Ideologies of Energy Development

Large-scale wood energy development has occurred in the context of a national and international debate about energy planning and policy. The larger debate and its connections with energy development controversies are curcial to understanding wood energy controversies.

In any ideological debate questions of epistemology and values inevitably crop up. Technical choices such as the one presently described for energy development are, therefore, as much political as scientific. Purely technical argumentation cannot bridge the gap between the viewpoints in the debate. The ideologies that distinguish proponents and opponents in technical debates rest on the values that each side wishes to preserve.

It is once again necessary to emphasize the importance of values at this juncture. If the technical disagreements in controversies were strictly about the facts, then they could be solved by technical experts rather than requiring a political solution. This treatment of the major ideologies in wood energy development controversies never wanders far from the value issues that invigorate these ideologies. Each ideological position belies an underlying basis in social or political values. Ideologies embody values and provide a prescription for their achievement. By adopting an ideology, an actor in a technical controversy acquires a program to guide his or her activities. Although this reckoning of ideology focuses on the study of wood energy development, there is reason to believe that it can be generalized to other development-environment controversies.[26]

The differences between small-scale and large-scale energy development have inspired a persistent ideological debate. The so-called "two paths" debate over future energy development appeared fully fledged following publication of a paper by environmentalist Amory Lovins in *Foreign Affairs*.[27] It lucidly defined the ideological differences between promoters of conventional energy development and those interested in conservation and alternative energy development. Extensive congressional hearings on the merits of the debate have served to point out the efficacy of arguments against headlong

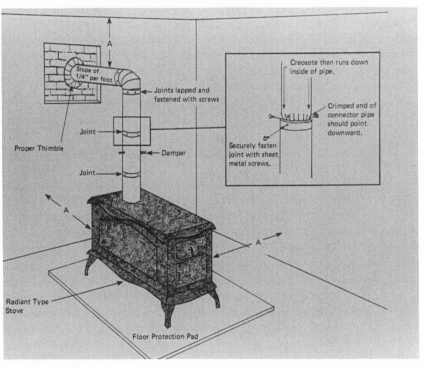

A WFPP (above) is a large-scale, centralized technology for the development of wood energy, by contrast with a wood stove (below) which is a small-scale, decentralized technology. *Picture below courtesy of University of Wisconsin-Extension.*

development of conventional energy technologies and to reinforce and diffuse the debate.[28] Widespread interest in this ideological debate has increasingly forced energy policy discussions into its mold.

Lovins argues that there are two mutually exclusive routes to solving the energy problem—the hard path versus the soft path. The hard path is characterized by large-scale, centralized, capital-intensive, technologically advanced supply systems. By contrast, the soft path relies on alternative energy sources and technologies that are small-scale, decentralized, not capital intensive, relatively simple, and environmentally benign. The United States and other industrial countries have been on the hard path since World War II. Policies such as subsidizing and promoting the use of nuclear power have engendered movement along this path. Lovins believes that the soft path is the correct policy choice. The transition to soft energy technologies that primarily utilize solar energy will, according to this view, entail energy conservation as well as aggressive development of renewable energy resources.

The touchstone for soft-path energy policy is to minimize adverse environmental effects. Those who advocate this policy undoubtedly draw the line at covering deserts with photoelectric cells or beaming power to earth from satellites. Decentralized, small-scale approaches like solar greenhouses and thermosiphon air panels are the preferred technologies for utilizing solar energy. They are arguably less disruptive of environmental values.

Large-scale development of wood for electric power would appear to be a litmus test for this policy choice in energy development. The choice could be posed this way: to burn wood for a low-grade purpose (space heating) at small-scale (the wood burner) and in a decentralized manner (in homes and buildings) *or* to burn wood to produce a high-grade energy form (electricity) in large-scale, centralized energy facilities (WFPPs). It should not be forgotten that there are other choices for utilizing wood and other alternatives for obtaining power. In any event, the hypothetical choice in wood energy development demonstrates the underlying ideological concerns. Economic interests come down on the side of continued expansion of energy production via the hard path, while environmental interests seek to preserve their values via the soft path. We shall soon see how this debate was played out in wood energy development controversies.

That development of wood for electric power is mainly a prospect for rural areas brings up an important consideration for the two-paths debate. Rural communities will bear the disproportionate

WFPPs constitute a large-scale development of wood for energy. Pictured are the turbines of the Pacific Oroville WFPP at Oroville, California. The combined rating of these turbines in 18.7 MW. *Courtesy of Pacific Energy.*

A sense of the scale of WFPPs is apparent in this photograph. Pictured is the control room of the 25.6-MW Chinese Station WFPP located at Chinese Camp, California. *Courtesy of Pacific Energy.*

burden of development. Casper and Wellstone state the case as follows:

> One clear distinction between the hard and soft energy paths is their impact on rural Americans. Farmers and other rural people are targeted to make the major sacrifices if we continue to move along the hard energy path. Soft energy strategies, with their emphasis on conservation, tend to distribute the sacrifices much more equitably between urban and rural people.[29]

Public opposition to proposals for large-scale development of energy resources is now routine. The essence of this opposition is concern for the maintenance of quality of life. The environmental movement has been instrumental in turning this sentiment into effective political action. It is well worth noting that public concern over the effects of large-scale energy technologies on environmental quality preceded the energy problem. Although wood energy seems to present risks and hazards of a much lower order of magnitude, centralized schemes for producing electric power with wood are also being carefully scrutinized for their environmental impacts. Ideology becomes the instrument for this scrutiny when the policy issues are debated.

4

A History of the Hersey Controversy

Michigan's Response to the Energy Crisis

The search for alternative energy sources as a substitute for a portion of the fossil fuels imported by Michigan—and for which dollars must be exported—began in earnest following the energy crisis of 1973–1974. Most of the action to assure secure supplies occurred at the national level. However, Michigan, along with many other states, took stock of the situation and proceeded to plan for a future of more volatile energy supplies. One result was the introduction of a generous renewable energy tax credit in 1979 to encourage the small-scale development of solar, wind, and low head hydro in Michigan.

The potential of another renewable energy resource soon headed the agenda—the abundant supply of wood biomass in the cut-over area of northern Michigan. The public, however, was already well ahead of government. Nationally, the number of wood stoves increased from one-quarter million in 1972 to two million in 1977.[1] By the 1983–1984 heating season, Michigan residents were consuming nearly six million six hundred thousand green tons of wood to heat their homes. Industry, mainly the forest products industry, used about one million one hundred thousand green tons for fuel, with utilities making use of between nine hundred and one thousand eight hundred green tons.[2] Overall about 85 percent of use of fuel wood was for residential space heating. The most notable response to the energy crisis in terms of wood fuel utilization was being made by individual consumers.

By the late 1970s wood from Michigan's extensive forests was perceived as a promising source of energy for something other than fueling a renaissance in space heating. The Michigan Public Service Commission surveyed the possibilities and concluded that northern Michigan was capable of producing more electricity from wood than that region consumed in 1976, and that proper management techniques could greatly expand the potential supply.[3] In November 1977 Michigan Governor William Milliken convened a national

Morbark Industries of Winn, Michigan became a venture partner in the Hersey WFPP. It hoped to market some of its wood chipping machines to supply fuel to the plant.

conference on wood energy development at the University of Michigan.[4] In a speech to the conference he suggested that a demonstrational WFPP be built in Michigan.[5]

Private interests responded to the challenge. Morbark Industries, Inc. of Winn, Michigan took a special interest in the proposal for it stood to gain financially by the expanded market for its whole-tree harvesting equipment. Parenthetically, Morbark was featured as a model of private initiative in energy development on then presidential candidate Ronald Reagan's national broadcast, a fact reported in a local newspaper at the beginning of the Hersey controversy.[6] Morbark Industries joined with Consumers Power company, the largest utility in the state (serving most of the state outside of the Detroit metropolitan area) and Wolverine Electric Cooperative, a rural electric cooperative based in Big Rapids, to undertake the venture. On 5 June 1978 they issued a special news release marking their agreement to pursue construction of a pilot electric generating plant in Michigan "to be fueled by wood and solid refuse."[7] This partnership subsequently contracted with Daverman and Associates, Inc. of Grand Rapids to do the feasibility study for a demon-

A demonstration wood chipper set up adjacent to the Morbark Industries plant in Winn, Michigan.

strational waste wood-fired generating plant that could have an ultimate rating of 50 MW.[8]

A network of groups was involved in planning the project. Members of committees and groups totaled forty-eight, including eighteen from Consumers Power (38 percent), seven from Morbark Industries (15 percent), six from Wolverine Electric Cooperative (13 percent), and seventeen (35 percent) from Daverman and Associates and W.P. London and Associates, the two consulting firms. The most influential individuals among project promoters were: J. N. Keen, manager of Wolverine Electric, and W. H. Sells of Morbark Industries, who served on four committees each; Peter Ratcliffe of Morbark Industries and A. J. Hodge, J. L. Schautt, and H. C. Wayman of Daverman and Associates, who served on three committees each.[9] With this elaborate organization, the benefit of expert advice, the blessing of the state of Michigan, and the cooperation of a rural cooperative utility and the state's largest utility, success for the venture seemed all but assured. The plant would be sited in a nonmetropolitan area where interest in local economic development presumably would be high.

Headquarters of Consumers Power Company in downtown Jackson, Michigan.

The Proposed Demonstrational WFPP

The demonstrational plant proposed for Michigan was to be a site specific 25-MW facility. It would provide enough electricity for twenty-five thousand to thirty thousand homes.[10] None of the technology involved in the power plant would be innovative. The utility partners decided not to take any risks in order to improve on the first generation hardware of WFPPs.[11]

The voluminous $300,000 feasibility study completed by Daverman and Associates of Grand Rapids estimated the annual fuel requirements for the WFPP at two hundred eighty-five thousand tons of green wood. About half (one hundred fifty thousand tons) was expected to come from state land. The forest base for the generating plant was put at one-half of 1 percent of the forest within a twenty-five-mile radius of the facility[12], a claim that became a lively source of technical debate. Enough wood was deemed available in the surrounding six counties to sustain the plant, although later a seventy-five-mile radius was established for wood to be auctioned from state land. Nevertheless, the feasibility study set the maximum economic distance for hauling wood chips at less than fifty miles.[13] Whole tree harvesting would be accomplished with the use of equipment manufactured by Morbark Industries. If completed by 1983 as planned, the WFPP would have been the first totally wood burning power plant in the country.[14]

Initially a 10–30 MW plant was considered for purposes of developing estimates of capital and operating costs. The analysis of scale established increasing economies with unit size, making a 30-MW unit the most economical choice within this range. The feasibility study concluded, however, that

> ... since the site selection and preliminary design are based on an ultimate plant size of 50-MW gross output, it was determined that a demonstrational plant will initially have a 25-MW unit which could ultimately be duplicated.[15]

Early Planning and Site Selection

The feasibility study looked at a wide range of factors in coming to a conclusion about the viability of the venture. The scope of activities for the study included:

- Determination of the adequacy and usability of a wood and/or a wood and refuse derived fuel (RDF) fuel supply
- Determination of fuel handling and ash-disposal methods

- Determination of equipment availability, optimal sizing, and conceptual design, including the determination of the mode of cooling and source of cooling water
- Production of schematic drawings of the fuel handling, boiler, and steam turbine of the wood and wood/RDF-burning electric generating plant
- Determination of the optimum locations for a wood-fired and a wood and RDF-fired plant
- Preliminary plan view of the major facilities of the plant
- Determination of the regulatory permits and approvals necessary for the entire project
- The appraisal of the likelihood of success in obtaining the necessary regulatory permits and approvals
- Determination of the estimated capital and operating costs of the electric generating plant
- Determination of the estimated cost of the design, and cost of the regulatory approvals and permits phase of the project
- Determination of financing options available to carry on the project and explore possible owner/operating arrangement[16]

The problem of building the WFPP was apparently viewed as largely technical and economic. If the plant was determined to be feasible on this basis then, the promoters assumed, it should and would be built.

Twelve sites were evaluated using the following selection criteria:

- Minimal environmental impact
- Location within service territories of Consumers Power and/or Wolverine Electric Cooperative
- Proximity to electric transmission facilities
- Transportation of fuel
- Adequate supply and availability of waste wood and/or RDF (for southern lower Michigan sites considered)
- Availability of sufficient land for development of a 50-MW site[17]

On this basis three northern lower Michigan sites were provisionally selected—Harlan, Hersey, and Whitehall. A special meeting was held in Big Rapids, a town about fifteen miles from Hersey, "to obtain input from individuals and organizations with particular environmental interests."[18] Public meetings were then held near each of the final three prospective sites during October 1978 in order to gauge public sentiment toward the proposal. The public meetings were summarized and evaluated as follows in the feasibility study:

In general, the public meetings appeared to establish overall public acceptance of the possibility of installing a waste wood-fired electric generating plant at each of the sites, provided that impact on the environment would be minimal. There was some opposition. There was also some apprehension as to possible odors at the Harlan and Hersey sites which might result if RDF is burned as a supplemental fuel.[19]

This summary of the meeting oversimplifies the concerns expressed at the hearings. Those who would soon serve as leaders and local experts for the opposition to the Hersey WFPP[20] honed in on the motives and methods of the proponents, stating, for example:

It seems to me like it's the same old thing. The large power companies are just willing to take all our resources, make their money and run off with them and leave us with a barred looking place [followed by applause].... What I'm concerned about is, once you get started on something like this, then it's hard to stop. Just like the nuclear power plants are hard to stop; once they get going on them, they've got all this money invested in them and they just continue.

And:

What I'm saying is the large power companies generally have more

An artist's conception of the Hersey WFPP released to the media by the venture partners. Compare with the McBain WFPP (p. 134). *Courtesy of Consumers Power Company.*

obligations to their stockholders than they do to the residents of the area where the plants exist.[21]

Hersey was soon selected as the proposed site "primarily on the basis that it will have, along with reasonable economics, the lowest environmental impact of all three sites."[22] Technical aspects of the Hersey site and of the WFPP hardware were analyzed and reported in the feasibility study. Perhaps as a strategy to counterbalance environmental concerns expressed at the site hearings, it touted a set of immediate socioeconomic benefits of installing the WFPP. These benefits are worth listing inasmuch as they present the developers' case for acceptance by the local community. The feasibility study enumerated the following seven socioeconomic benefits of the project:

- The impact on local merchants and service industries during construction when the labor force could approach two hundred workers
- A resulting increased employment opportunity for construction trades in western and central Michigan during construction
- A permanent increase of about eighty new jobs represented by the plant staff and wood gathering personnel
- Increased income to private property owners through sale of waste wood from what are now marginally productive woodlands lacking a market
- Increased forest yields and improvements to wild game habitats resulting from selective harvesting of waste woods
- Additional revenues to the state of Michigan from the sale of waste woods under its forest management program, waste wood that the state must now pay to have removed
- A demonstration to the remainder of the Michigan electric utility industry of the benefits to be derived by the concerted efforts of the private and consumer-owned sectors of the industry in conjunction with private industry accomplished without the expenditure of governmental funds[23]

Final recommendations of the study suggested authorization of final design, financing, and construction, contingent on wood supply commitment from the Michigan Natural Resources Commission. The response from the Hersey community proved to be less enthusiastic than from the venture partners who had commissioned the feasibility study, as will soon be revealed.

To map out the initial reception in Hersey, it is necessary to digress to the period before it was selected as as site. The first *Osceola County Herald* account of the proposal summarized the first public hearing on the project held in Big Rapids.[24] The two issues upon which the controversy would hinge—impact on area forests of large-scale harvesting and the burning of solid waste in the WFPP—were evident in that account. This 28 September 1978 article carried the semi-accurate headline "Hersey Site Chosen for Chipping Plant." Final selection would not be made until February 1979.

During September 1978 hearings were held at or near each of the three prime sites selected—Harlan, Hersey, and Whitehall. Citizens of Hersey expressed reservations at their hearing in Reed City and subsequently organized to explore alternative viewpoints on the project. A study group formed to gather information about the impact of the project on the Hersey community. When Hersey was announced as the site concurrently with release of the Daverman and Associates feasibility study in February 1979, the conflict was already well established. The Committee for the Rational Use of Our Forests (CRUF) was organized late in December 1978 out of that community study group. The leaders of CRUF were relative newcomers to the area. They were seeking to protect the environmental values that had attracted them to the region. Despite their status as newcomers, action by CRUF members soon generated a rising tide of community opposition to the WFPP.

A Nonmetropolitan
Community Confronts Technological Change

The fact that Hersey is rural and remote from population centers did not portend the ensuing controversy. Hersey is the sort of town one would expect to welcome such a project. Until recent decades such communities usually lacked the kind of diversity that gives rise to citizen protest. The appeal of additional jobs and growth in tax revenues would seem to be the paramount interest in an economically disadvantaged community like Hersey. Ironically, its location among scenic and recreational resources is essential to understanding the community response to the WFPP proposal. The population migration turnaround—a change clearly connected to these amenities—greatly increased pluralism in the community. The resulting diversity of ideologies and political skills formed the basis for a successful opposition movement.

To understand the response of the Hersey community to the prospect of a large-scale wood burning power facility, it is thus necessary to look at the environment in the area and how residents, old-timers as well as newcomers, valued it. This account begins with

Figure 2. Location of Hersey and Indian River, Michigan

a description of the community and its experience with wood fuel, followed by a discussion of important social changes that led to controversy.

Hersey is located in Osceola County, which lies within a scenic and recreational region in the northern half of Michigan's lower peninsula (see figure 2). Manistee National Forest borders Osceola County to the west, while Chippewa State Forest occupies much of its central portion. Numerous lakes, streams, and rivers are found within its boundaries. Three major highways provide easy access to and from any part of the state. The nearest major city is Grand Rapids some seventy-five miles directly south.

Hersey township has the longest history of settlement in county. Hersey village was the county's first cental place. In this study the Hersey community is identified with the township. Political activity in the controversy occurred largely at this level since the proposed site was in the township three-quarters of a mile southeast of Hersey village.

The Hersey area had long enjoyed the forest as a scenic and recreational amenity. As a result of the energy crisis many county residents again exploited the forest for fuel to heat their homes. This

Downtown Hersey in 1979 (above) and 1989 (below). As you can see, it did not change much in appearance during the last decade, notwithstanding growth in the surrounding area due to inmigration. *Picture above Courtesy of Bernadette Miller.*

TABLE 1

**Population Trends in Hersey Village,
Hersey Township, and Osceola County, 1950 to 1980**

Geographic Unit	Population 1950	1960	Percentage Change	Population 1970	Percentage Change	Population 1980	Percentage Change
Hersey Village	239	246	2.9	276	12.2	364	31.9
Hersey Township	282	399	41.5	539	35.1	1,299	128.0
Osceola County	13,797	13,595	−1.5	14,838	9.1	18,928	27.6

Source: U.S. Bureau of Census, Census of Population, 1950, 1960, 1970, 1980.

familiarity with wood burning and the low density of settlement suggested to residents that the use of wood as fuel was not an environmental problem, at least for residential wood burning. The WFPP came to be seen as a radical departure from this experience. The threat to the forest resource itself and the prospect of burning municipal waste (RDF) in the power plant fomented opposition. Notwithstanding familiarity with wood as fuel and the assurances of the feasibility study, the Hersey WFPP would prove controversial for the community.

Production of steam and/or the generation of electric power using wood fuel was not novel for the community either. The practice was common in the United States early in the century, even in Osceola County. One elderly resident (who, incidentally, became an opponent of the WFPP) reported that his grandfather operated a wood burning power plant in Evart—a town just to the northeast of Hersey—in the early 1900s.[25]

An examination of the population migration turnaround in the county and Hersey Township implies a connection between population change, environmental values, and controversy. Osceola County contains only six towns and villages, all of which reported populations under two thousand five hundred in 1980. During the 1970s its population virtually exploded, rising from 14,838 in 1970 to 18,929 in 1980 (refer to table 1). The gain of 4,090 equates with a growth rate of 27.6 percent. Most of this growth was due to in-migration. Newcomers generally settled in open country areas of the county (refer to table 2). Only 6 percent of this growth accrued to the towns and villages. The overwhelming proportion (94 percent) occurred outside existing population centers. In-migrants to the county were in the vanguard of those repopulating the countryside. The experience of the population migration turnaround here is

TABLE 2

Growth in Towns and Villages Compared to that of Open Country for Hersey Township and Osceola County, 1950 to 1980

Geographic Unit	Population Change 1950–1960	Percentage	Population Change 1960–1970	Percentage	Population Change 1970–1980	Percentage
Hersey Township, excluding Hersey Village	110	94.0	110	78.6	602	87.2
Hersey Village	7	6.0	30	21.4	88	12.8
Hersey Township, Totals	117	100.0	140	100.0	690	100.0
Osceola County, excluding towns and villages	−411	—	1233	98.4	3825	93.5
Osceola County	209	—	20	1.6	265	6.5
Osceola County, Totals	−202	—	1243	100.0	4090	100.0

Source: U.S. Bureau of Census, Census of Population, 1950, 1960, 1970, 1980.

typical of that in many remote nonmetropolitan areas of the United States.

Findings of a study of housing growth in Osceola County generally confirm those of previous studies examining the importance of environmental and locational features in the growth of population in nonmetropolitan areas.[26] The attraction of water seems to be compelling. The type of water body (lake or river) apparently does not matter. Forests are probably more influential than actually revealed by this study since only public forestland was considered whereas much of the private property in the county is also forested. Highway access was also found to be an important predictor of growth. Roadways seem to facilitate open county settlement. Road access also affects the type of housing constructed, for example, the association between mobile homes and highways.[27]

This study of housing growth used county health department data rather than census data and independently found that Osceola County grew in the 1970s largely because of the population migration turnaround. Growth during this period proved to be volatile. New construction declined by over 50 percent during the nine-year study period (1970–78). Such growth can easily take on the character of a boomtown. In the boomtown scenario growth is brought on by the sudden intensification of usage of a local resource. For areas like Osceola County, this resource is the environment itself.

As the more desirable sites are consumed, growth spills over into previously less valued sites. Later growth thus involves a "filling in" process in areas between more remote places and population centers, but as usual the resource is finite. Eventually additional growth cannot be accommodated without altering the environmental amenities that precipitated it. Scarcity of available prime sites also drives up the cost of land and thus redefines the market. Consequently, desirable sites may be priced out of the reach of locals and wind up in vacation or retirement homes for more well-to-do migrants.

The population turnaround in Osceola County is apparent in the plat map for the area. Extensive subdivision along the Muskegon River within the township is an obvious feature of the map.

Environmental values evidently stimulated growth in Hersey Township much earlier than in other parts of the county and were responsible for disproportionate growth there in the 1960s and 1970s (tables 1 and 2). Like the county as a whole, most of the recent growth in Hersey Township has occurred in the open country. Although many of those who returned to the county were retirees, an important component of the turnaround has been the young and well-educated. One factor promulgating in-migration has been the growth of small-scale manufacturing. Increased job opportunities also helped to retain the county's crop of youth. About four-fifths of the migration stream to Osceola County can be attributed to urbanites—most from downstate central cities, especially Detroit and Grand Rapids.[28]

From a sociological perspective this case study is all the more interesting because it involves a nonmetropolitan community. Protest against such facilities more often arises in urban communities.[29] Environmental amenities surrounding Hersey had attracted newcomers to the area. Scenery and recreational possibilities were foremost among these. With these migrants came new values and ideologies, significantly increasing pluralism in local politics. This change is reflected in the character of the Hersey plant opposition leaders—young, well-educated, and recently introduced to the area from urban areas to the south. They showed a commitment to preserving the values that had attracted them to Hersey, and cooperated with locals to accomplish their mutually desired goals.

In a parallel study of opposition to coal-fired power plants,

Aldrich notes the strong influence of the population migration turnaround in the local politics of siting. He uses the term "local expert" to describe the technical expertise utilized by opponents to fight proposed power plants. As in the Hersey case, local experts turned out to be recent migrants from urban areas.[30]

Demographic changes are an indirect cause of the controversy that engulfed the proposed WFPP. The influx of newcomers has entailed qualitative changes that are often not addressed in assessments of the impact of nonmetropolitan growth. Such qualitative changes in local politics are evident in the northern lower Michigan region. Analyses of the turnaround for the community in question as well as the northern Michigan region are reported in detail in other studies.[31] The purpose here is to look beyond quantitative details of the turnaround to a fuller appreciation of its multifaceted effects in a technical controversy. After all, the values, ideologies, and political skills brought to the area by newcomers proved to be the basic ingredients of the controversy.

The Initial Local Response

The public meetings had revealed enough about the project to give citizens, who conceivably might see things differently than the experts from Daverman and Associates, the opportunity to consider a response. Although Hersey was not the unequivocal site until the feasibility study was released four months later, the fact that the community was in the running proved sufficient to arouse some Hersey area residents. By the time a meeting was held on the Hersey site a relatively large group of citizens took an interest in the proposal. "Approximately 100 people, mostly from the Hersey area, attended the fourth and final public hearing concerning the plant at the Osceola Inn [Reed City]." The *Osceola County Herald* headline tells the story: "Opinions Split on Plant Near Hersey."[32] Three weeks later Hersey village President Bion Jacobs, a lifetime resident of the area, convened a public meeting in the Hersey township hall for the purpose of letting local citizens express their views on the proposed WFPP.[33] In doing so, he responded to a petition which asserted that the issue had not been fully aired in Hersey. Approximately fifty people attended that meeting. The newspaper account documents the formal response of the Hersey community:

> The outcome of . . . the meeting was the appointment of a four member committee to "keep on top of the situation and to investigate the matter

Hersey WFPP activist Marco Menezes and his daughter, Sunshine, pose for a picture published with an article on the controversy in *Michigan Natural Resources* in 1979. As an effective leader of the opposition, Menezes defended community values threatened by outside interests. *Courtesy of Bernadette Miller.*

further." Appointed to the committee were Fred Cole, Dave Springer, Ken Ford, and Marco Menezes.[34]

Interestingly enough all were relative newcomers to the area. They would soon become the leaders of the opposition, using their organizational skills to protect local resources based on a view of resource development at odds with that of the venture participants.

But at the outset native member of the community were sowing the seeds of discontent about the way the project was being planned. During the public meeting Jacobs articulated the doubts of many in the community: "It was pretty one-sided in Reed City. They gave some very vague answers. You'd think they'd have some concrete answers if they're going to spend that kind of money."[35]

One could easily assume from this statement that the dispute was about to be thrashed out on the technical aspects of the project. Jacobs then made the most quotable statement of the controversy: "When I first heard about the plant I thought it was the greatest idea, but I'm getting tired of reading in the newspapers how other people are saying that they want the plant in Hersey. They don't live here. We do. Don't we have any say? We want to make our own decision."[36] The issue of the threat to local autonomy represented by the WFPP had been joined. Jacobs's viewpoint reflected a NIMBY response by many in the community, but the leaders of the opposition would have different ideas. They would focus on the threat to the forests and use whatever issues in the technical debate that might help defeat the project.

Members of the committee formed at that first Hersey meeting gathered at Henry's Bar in Reed City several weeks later to discuss the problem. Over drinks they decided to initiate an opposition group, agreeing on the name Committee for the Rational Use of Our Forests—CRUF for short. During the next two years CRUF would incorporate, solicit funds, aid other opposition groups in the region, and oppose the Hersey WFPP in every available forum. CRUF was composed mainly of relative newcomers to the area who possessed political skills, environmental interests, and some technical expertise related to managing forest resources. A rising tide of opposition, spearheaded by CRUF, awaited the decision to site the WFPP at Hersey. Several articles in the *OCH* favorable to wood-electric power and detailing the Burlington, Vermont experience did not quell community apprehension.[37]

When the feasibility study was released and the Hersey site announced in early February 1979, the opposition was well established. A meeting held in Hersey Elementary School on 6 February 1979 was reported even in the distant Detroit newspapers. That meeting overflowed the facility and was estimated at two hundred to two hundred fifty people. The *Detroit Free Press* article assessed the situation as follows:

> Most of the audience showed some hostility to the project. Their attitude seemed to be that although the corporate representatives and State

Department of Natural Resources officials had plenty of facts and figures, a citizen just can't trust government or big business anymore.[38]

The journalist's conclusion in this article is reinforced by the statement of a Hersey resident who was moved to write a letter to *Michigan Out-of-Doors* magazine:

> ... [at the] meeting ... held in Hersey ... a number of knowledgeable men representing the involved companies spouted rhetoric blocking opposition and questions with figures and statement which a person of average intelligence could not dispute.... These are companies and individuals who will gain monetarily in this venture and then leave, while those of us whose needs and life-styles are less complex will be left to pick our way among the stubble and worry about soil conditions. As many in this area heat by wood, we will not longer be able to take our families into the woods to gather firewood.[39]

During the four hour meeting citizens repeatedly expressed concern about the issues of burning solid waste in the WFPP and adverse effects of clear-cutting forestland.[40]

Walt Grysko, a feisty local columnist, got into the conflict on the side of the opposition. His first column on the dispute—"Hersey Woodchip Plant Monster!"—marked a change in the style of the debate.[41] He questioned the technical claims made on behalf of the WFPP and the validity of comparisons with Burlington, Vermont. The intense debate at this early stage focused on technical matters despite the evident value preferences.

The Ebb and Flow of Political Activity

A detailed chronology of the Hersey controversy is provided in Appendix 1. A number of its salient events are elaborated in the rest of this case history.

By the end of February the *Osceola County Herald* was extensively reporting the contrasting views of promoters and opponents. On 26 March 1979, CRUF released a well documented position paper attacking the feasibility study.[42] The Daverman and Associates study had cost $300,000. By contrast, CRUF leaders had volunteered their time and expertise in forestry, engineering, and business to draft a twenty-one-page rebuttal. These local experts undertook research on technical issues related to their concerns. The position paper reached the conclusion that the plant was not feasible for

The Osceola County Herald located in Reed City became a lightning rod for the Hersey WFPP controversy. The content of articles on the controversy was analyzed for purposes of this study (see chapter 7).

environmental reasons and, therefore, should not be built. Knowledge of the technical aspects of wood-electric power and large-scale wood harvesting, a dedication to stopping the WFPP on technical grounds, and a capacity for research and technical writing are all apparent in their report.

The very existence of two feasibility studies, even though one was far more slick and sophisticated than the other, created controversy. The promoters were forced to respond at length to the CRUF challenge. In retrospect it is clear that the public, at least the citizenry of Hersey township, found CRUF's analysis to be credible. CRUF rapidly gained legitimacy in the community to match that of the experts from Daverman and Associates and the Michigan DNR. The attempts by project promoters to respond to a barrage of criticism did not reduce public concern or mitigate the effectiveness of CRUF. Indeed, the opposition seemed to gain ground as debate increased.

The initial period of intense debate in the media and at public hearings culminated in an extensive report in the *Osceola County Herald* written by two reporters from the *Big Rapids Pioneer*, one of whom would soon be hired as editor of the *Osceola County Herald*. The report was preceded by an editor's note setting out its purpose, that is, to better present the issues involved. Each reporter acted independently in interviewing and summarizing the views of proponents and opponents of the WFPP. The next critical step was noted in the article—the long-term commitment of wood supplies by the state. The incipient debate in "The Wood Chip Controversy" was appropriately subtitled "Pro: Money's a Key" and "Con: Use Is Inefficient."[43] The promoters interviewed emphasized the WFPP's contribution to the tax base, while the opponents questioned Daverman and Associates' figures on its technical efficiency. Ken Ford summed up the case against, pointing out that " . . . the plant is the least we can get for this wood. Let's aim for something higher."[44]

A four month lull in the controversy preceded deliberations by the Michigan Natural Resources Commission on wood supply commitment. CRUF revised its strategy of stopping the plant in this forum, soliciting an environmental lawyer in Traverse City in July 1979 for the purpose of drafting a waste ordinance for Hersey township. At the same time CRUF actively sought out other forums for opposing the WFPP, for example, the wood energy conference sponsored by the Michigan Forest Association held in Lansing 2–3 November 1979. CRUF argued against committing wood from public lands at the meeting of the NRC on 6 September 1979 at Higgins Lake. At first it appeared that the decision would have to await a protracted environmental assessment.[45] However, on 12 October 1979, a "slightly hesitant" NRC agreed to auction off enough excess state grown timber to fuel the plant during the next ten years.[46]

CRUF held a press conference at the Hersey Township Hall on 17 October 1979. Although unhappy with the NRC decision, the group did not concede defeat. CRUF representatives stated that they recognized who they were up against—"the DNR and Consumers Power, two of the most powerful units in the state"—but that they nevertheless put their chances of defeating the project at 50–75 percent. CRUF leaders complained that despite the technical data they had presented, the NRC had failed to grant them the recognition [legitimacy] they deserved.[47]

Although the NRC apparently took a skeptical view of CRUF, the township of Hersey had already formed a different opinion. The frustrating experience in dealing with state agencies was a catalyst for ensuing political action in the local community. CRUF sought help in this arena well before the NRC decision. One of the experts for the CRUF put the transition in strategy this way:

We urge all local, township, and county officials to consider adopting similar ordinances[48] to protect themselves from imported environmental degradation. Experience has shown that the most effective, responsive form of government is at the local level. We in rural Michigan will do whatever is necessary to strengthen our hand against Lansing, which has clearly thrown us to the wolves.[49]

The focus of interest, then, quickly shifted from forest impacts— the domain of state regulation—to local environmental effects—a domain for local regulation. The waste issue moved front and center. Toxic, as well as municipal, waste would now be the source of debate. A petition expressing concern about the prospect of garbage being burned in the plant was presented to the Hersey Township Board on 16 October 1979. It contained three hundred thirty signatures, an indication of widespread opposition in the Hersey community. The possibility that the plant might ultimately be rated at 50 MW and the specter of five hundred tons of RDF per day coming up from Grand Rapids prompted columnist Grysko to predict that the WFPP ''. . . could even conceivably make Osceola County the state trash disposal area or the state dump.''[50] The petition called for the adoption of an ordinance to prohibit the burning of solid wastes in the plant. The stated plan of the promoters was for a ninety day test burn of RDF. The petitioners feared that this would only be the beginning.

Walt Grysko and John Keen of Wolverine Electric exchanged barbs in the next several issues of the *Osceola County Herald*. Grysko questioned the claims about jobs, recapitulated the RDF issue, and brought up the matter of local fiscal costs associated with such a development. He urged county commissioners to restrict the burning of RDF in the county. Keen denied Grysko's earlier claim that the plant might eventually be rated at 50 MW. He admitted that RDF would be used in the WFPP, but only on an experimental basis and for no longer than ninety days. Keen asserted that the test was required to determine the effect of RDF on plant equipment. He even suggested that ''. . . perhaps Mr. Grysko would . . . consent to letting us utilize some of the tripe that he served up in his October 18 article as a portion of the RDF.''[51] Claiming that the opposition had overreacted and trumped up charges against the plant, proponent Keen invited *Osceola County Herald* readers ''who would like to find out *factual* information'' [emphasis mine] about the WFPP to contact him.

Initially, CRUF activity and media accounts of the controversy emphasized the prospect of damage to the forests (in particular nutrient loss due to whole tree harvesting) and possible environmental deterioration in the areas surrounding the site for the WFPP. During the first year of the conflict CRUF had sought to stop the

The prospect that municipal refuse might be shipped from Grand Rapids by rail car and burned in the Hersey WFPP became an issue in the controversy. Eventually Kent County built a cogeneration facility to burn some of the solid waste produced by Grand Rapids. Two views of the facility are pictured. It opened in 1989.

plant by lobbying the Michigan Natural Resources Commission to veto the plan to allocate wood from state forest land as requested by the promoters. When this approach failed, CRUF focused on the issue of burning refuse derived fuel (RDF) in the WFPP. The rest of the controversy reflects the effectiveness of this tactic.

The luck of the promoters took a turn for the worse despite the NRC decision in their favor. On 13 November 1979 the Hersey Township Board unanimously adopted the Solid Waste and Toxic or Hazardous Substances Disposal Ordinance. The new ordinance empowered the township to regulate transportation, burial, or burning of wastes within its jurisdiction.

The tempo of the controversy increased after this decision. The entire concept was at this point called into question when, on 15 November 1979, the *Osceola County Herald* gave front-page billing to the news that Burlington, Vermont had shelved plans to build a 50-MW WFPP and that instead it would buy power from Ontario Hydro.

CRUF continued to oppose the power plant at every opportunity. At a 10 December 1979 meeting of the Osceola County Board of Commissioners, CRUF lobbied for a legal document assuring that wood chips would be the primary source of fuel in the WFPP. The three hour session was well attended and included representatives from the two utilities, the DNR, and Hersey residents and officials.

This meeting yielded a concise description of the actors in the controversy. Walt Grysko made the following observation in his column. The nature of the disputants and some indication of their values are apparent in this report.

> ... looking over the group, an impartial observer could not help but notice that the opponents to the Hersey plant were with one exception all young people, who would gain very little, it anything, financially from the new plant's construction. These people seemed to worry about the ecological effect this venture would leave on the area and it's [sic] health effect on the Hersey township residents. Many of them have studied the "Feasibility Study"[52] which mentions a few of the uncertainties and some of the health hazards with which they are concerned. On the other hand all of the outspoken proponents were older people, who seemed to think that this plant was a financial cornucopia; that it would help the area tax-wise, would provide new jobs, and would help them sell their wood products at a greater profit.[53]

The controversy was now down to just one issue—the burning of RDF in the plant. Pursuant to the exhortations of the township lawyer and a challenge by officials from Consumers Power to rescind or amend the ordinance, the township acted in a conciliatory manner. On 20 December 1979, Hersey township endorsed the plant and proposed an amendment to the ordinance adopted a month earlier. Those attending the meeting called for a referendum on the question of whether or not the citizens of Hersey township also endorsed the WFPP. Several attending that meeting took the opposite stance, holding that the Hersey Township Board should rescind the ordinance. However, this was clearly a minority position.

The proposed plant was ranked as the third most important story in the county during 1979 by the *Osceola County Herald* in its first issue of 1980.[54] The climax was yet to come.

A public meeting was held on 10 January 1980 to deal with what were described by Bion Jacobs as scare tactics by Consumers Power. The ordinance had been drawn up by the "best environmental lawyer in the state." It was the sixth such ordinance the attorney had drafted. Opponents of the WFPP were therefore skeptical of the criticisms leveled against the ordinance by the project promoters. Moreover, because the state Hazardous Waste Management Act went into effect on 1 January 1980, any local ordinance on the books before that date stood a much better chance of holding up in court. Discussion at the meeting consequently revolved around amending rather than rescinding the ordinance. A conciliatory meeting of the attorneys for all concerned parties that took place several weeks earlier in Traverse City produced a compromise amendment. The Hersey Township Board voted to amend the ordinance, exempting certain categories of operations from having to comply with the otherwise strict law. The amendment also allowed the utilities to apply to the board for a permit to burn RDF for a short period of time.

The controversy seemed to be over until March 1980 when the utilities applied for a permit five years before the test burn would take place. Concern increased among opponents that RDF (or even toxic waste) might become regular fare for the WFPP. Hersey township officials postponed a decision on the application at a meeting on 18 March 1980, notwithstanding the claim of the township attorney that Consumers Power might back out if the contract were not signed immediately. A public hearing on the matter was scheduled for 10 April 1980.

The death knell for the WFPP was sounded at the April 10th hearing. The Hersey Township Board recognized the groundswell of public disenchantment with the project. Rather than risk a decision in the emotion charged atmosphere prompted by the test burn application, the Hersey Township Board unanimously voted for a referendum on the question of granting the utilities a permit. "Tempers flared a bit on both sides" at the meeting, to quote the *Osceola County Herald* account. One opponent toted a bag of what he believed might be in RDF. Consumers Power officials had their own sample to pass around for inspection. Disturbances ensued when the possibility of burning PBB-contaminated carcasses in the WFPP was mentioned. Township Supervisor Forest Benzing threatened several times to end the meeting if the disturbances persisted. When Consumers Power attorney Jack Shumate said the site would be reevaluated if the test burn permit was not approved, applause erupted. One resident interjected: "what you want to do isn't the only thing that needs to be done. We need to live."[55]

Having failed in their attempt to budge the township on the compliance contract referendum, the utilities announced they would indeed reevaluate the site.[56] Some in the media viewed this result differently as suggested by the headline in the *Detroit News*—"Tiny Town Beats Power Plant Bid by Two Utilites."[57] The decision of the utilities was reportedly based on their need for "an orderly planning process aimed at minimizing delays and uncertainties." The referendum apparently was the straw that broke the camel's back. For the next several months the *Osceola County Herald* served as a forum for numerous recriminations from both sides in the controversy. In the atmosphere of mistrust between the Hersey community and the utilities, and in view of the very successful petition drive calling for the ordinance in the first place, utility officials had little cause to hope for approval of the contract. In May the Hersey Township Board voted to withdraw the referendum but resolved to reinstate it automatically if the utility companies should decide to go ahead with the venture. CRUF leader Doug Miller confidently asserted that the WFPP would never be sited at Hersey, but not because of the ordinance. Rather he suggested that both the community and the utilities now knew that the wood supply would not be adequate for the plant. Despite its strategy of defeating the plant through a waste ordinance, CRUF continued to emphasize forest resource issues. At about this time another suspicion was forming in the community regarding the use of the WFPP—that the desire of the Michigan DNR to have a facility for disposing of the state's toxic wastes was left unstated.[58]

The utilities withdrew consideration of the Hersey site in May 1980. They had apparently weighed the situation and concluded that the community would not approve the permit or eventually welcome the WFPP.

Citing lack of capital due to the economic recession, Wolverine withdrew from the venture in September 1980. Almost simultaneously Consumers Power announced that it would put off further consideration of the project for two to three years.[59]

The effect on politics in Hersey extended well beyond the decline of the proposal. The November 1980 election produced a watershed of changes in the leadership of Hersey township. Of the four candidates elected to the Township Board, two got there with active campaign help from area residents who had participated in CRUF. A very significant change was replacement of Township Supervisor Forest Benzing, who had been perceived as a proponent of the WFPP, by Bion Jacobs in a very close vote. Bion Jacobs, formerly the president of Hersey village, was known as a staunch opponent of the WFPP. He was one of the candidates who received campaign help from former CRUF members. In a race for the Osceola County Commission, a former CRUF member was narrowly defeated despite carrying Hersey township by a wide margin.

Issues in the Hersey Controversy

The following analysis of issues in the Hersey controversy is a spinoff of the content analysis summarized in chapter 7. It utilizes the public debate in the *Osceola County Herald* and transcripts of the hearing for the Hersey site. Issue categories were not prejudged. Rather, as arguments were coded, a set of specific issue categories emerged. In this fashion they "fell out" of the analysis. Iteration insured that an adequate and complete set of categories was generated. Below are listed the issues that were coded and analyzed. For purposes of the tables that follow, issues are presented by an identifying term followed by a brief description:

- *Corporations*—abuses of corporate power
- *Economics*—economic benefits or detriments to the area
- *Energetics*—energetic efficiency of the WFPP
- *Energy Crisis*—energy problem in general
- *Forest Ecology*—ecological impact, especially on forest soils and wildlife
- *Forests*—decimation or enhancement of the forest resource
- *Local Autonomy*—local control, protection of way of life
- *Local Environment*—impacts on the local environment
- *Renewability*—wood as a renewable resource
- *Scale*—scale and centralization in energy resource development

- *Social Equity*—equity in allocation of wood for area loggers and home wood burning
- *Solid Waste*—prospective use for and health/environmental effects of burning solid waste
- *Toxics*—prospective use for and health/environmental effects of burning toxic wastes
- *Waste Ordinance*—waste ordinance adopted by Hersey Township in October 1979
- *Waste Wood*—utilization of unmerchantable wood, a resource argued to be otherwise wasted
- *Wood Supply*—adequacy of wood supply for the WFPP

The Hersey site hearing was a harbinger of things to come. Table 3 shows that scale of development, solid waste burning, impacts on the local environment, and ecological impact on the forests were already concerns at the outset. The analysis casts a finer net than the conclusions about public concerns presented in the prime site hearings (see table 4). As will shortly be discussed, the waste issue occupied a similar prominence in the debate as a whole. The energy problem in general, economic benefits/detriments to the area, and

TABLE 3

**Issue Analysis of Arguments in the Hersey Site
Hearing Held 24 October 1978**

Issue	Rank	By Proponents	By Opponents	Total Arguments
Solid Waste	1	14	6	20
Scale	1	9	11	20
Forest Ecology	2	8	6	14
Local Environment	2	8	6	14
Economics	3	11	0	11
Energy Crisis	3	6	5	11
Wood Supply	4	5	5	10
Local Autonomy	5	3	5	8
Social Equity	5	4	4	8
Forests	5	4	4	8
Waste Wood	6	7	0	7
Corporations	7	0	5	5
Energetics	8	1	1	2
Renewability	8	2	0	2
Waste Ordinance	—	0	0	0
Toxics	—	0	0	0
Totals	—	82	57	139

adequacy of wood supply for the WFPP were of marginally lower concern in the hearing. Although prominent in both the hearing and the *Osceola County Herald* debate, most of these issues would decline in relative importance. However, adequacy of wood supply and benefits/detriments to the area economy, along with the overriding issue of waste burning, remained high priorities throughout the controversy.

TABLE 4

Feasibility Study Conclusions about Prime Site Hearings

A. The environment must be protected. Concern was expressed not only for air and water quality, aesthetics, and the impact of trucking at the proposed plant site, but also the possible effects of wood harvesting on the forests and wildlife therein.

B. The concept of wood as a source of electric energy was generally acceptable although there were objections.

C. The supplemental burning of refuse derived fuel (RDF) was generally opposed. With only one or two exceptions, general opposition was expressed to the burning of RDF at all but the Whitehall site. The panel members were quite candid in their responses to the audiences. It was the general opinion of the panel that a plant burning RDF from outside the area would meet substantial opposition. However, at the Whitehall meeting there appeared to be no major objection to burning RDF regardless of the source.

D. The method of forest harvesting was a concern. Different individuals were concerned over indiscriminate forest clearing and expressed a desire for some type of responsible forest management.

E. The additional income to the community that would result from employment associated with the plant and from wood harvesting was appreciated.

F. There were conflicting opinions relative to the aesthetic effects and future growth in managed forests using the types of wood harvesting discussed in this Report.

G. Apprehension was expressed regarding the effect selective harvesting for a wood plant might have on DNR permitted removal of fuel woods from state forests.

Source: Daverman and Associates, Inc., *A Feasibility Study for a Waste-Wood Electric Generating Plant*, prepared for Consumers Power Company, Morbark Industries, Inc., and Wolverine Electric Cooperative, Grand Rapids, Michigan, February 1979.

The salient issues in the controversy as a whole and a strong indication of their dynamics is shown in table 5. The conflict at first focused on area economic impact, effects of whole tree harvesting on forests, and adequacy of wood supply for the plant. As the controversy developed, the emphasis shifted to the burning of solid waste and to local autonomy. Area economic benefits/detriments and solid waste burning were persistent concerns in the controversy, the latter increasing in relative importance to become the single most important issue. The trend is especially pronounced if the late breaking issues of toxic waste and the waste ordinance passed by Hersey township are aggregated with waste burning.

TABLE 5

Ranking of Issues in the Hersey Controversy

Issue	Total	Rank	Phase I*	Rank	Phase II*	Rank
Economics	96	1	64	1	32	3
Solid Waste	80	2	21	5	59	1
Wood Supply	64	3	35	3	29	4
Forests	62	4	48	2	14	8
Local Autonomy	54	5	16	6	38	2
Forest Ecology	42	6	26	4	16	6
Social Equity	31	7	13	9	18	5
Local Environment	29	8	16	6	13	9
Waste Ordinance	29	8	—	—	29	4
Energy Crisis	21	9	14	8	7	12
Energetics	20	10	7	10	13	9
Waste Wood	20	10	14	8	6	12
Renewability	18	11	15	7	3	13
Toxics	15	12	0	—	15	7
Scale	12	13	3	11	9	10
Corporations	11	14	3	11	8	11
Totals	604	—	300	—	304	—

*A natural break in the controversy that occurred in May–June 1979 was used for purposes of dividing the controversy into two distinct phases. Phase I is from September 1978 through June 1979; Phase II begins July 1979 and ends September 1980.

The rise of value issues and the growing emphasis on local autonomy issues is supported by a further analysis of the debate summarized in table 6. This table is based on an aggregation of issues. Issues were placed into three major categories for purposes of the analysis—economic, environmental, and social/political. The

TABLE 6

Temporal Distribution of Economic, Environmental, and Social/Political Issues in the Hersey Controversy

	Proponents			Opponents		
	Phase I *	*Phase II* *	*Total*	*Phase I*	*Phase II*	*Total*
Economic	78	22	100	48	52	100
Environmental	36	64	100	30	70	100
Social/Political	69	31	100	22	78	100
Total	64	36	100	35	65	100

	Totals		
	Phase I	*Phase II*	*Total*
Economic	64	36	100
Environmental	33	67	100
Social/Political	33	67	100
Total	48	52	100

*A natural break in the controversy that occurred in May–June 1979 was used for purposes of dividing the controversy into two distinct phases. Phase I is from September 1978 through June 1979; Phase II beings July 1979 and ends September 1980.

distribution for these categories are expressed as percentages. Note that the N in table 6 is greater than for issue arguments coded in table 3 because some issues are double counted for purposes of this classification.

The economic, environmental, and social/political issues enumerated in table 6 are distributed in a ratio close to 3:2:1 for the controversy as a whole. Overall the opponents carried more of the debate in terms of the major issues (327 to 277). Except on economic issues—where proponents have the edge by nearly two to one, the debate was fairly even during Phase I. Opponents utilized social/political issues in Phase II, while proponents modestly shifted toward environmental issues.

Not surprisingly opponents were less prepared to mount a response during Phase I. In purely quantitative terms they were out-debated 180 to 120 or on a ratio of three to two. However, the opposition reversed the advantage during Phase II when the figures changed to 207 to 96 in favor of opponents, an edge of more than two to one. Opponents emphasized economic issues in Phase I. Environmental and social/political issues became almost twice as prominent in Phase II, with economic issues declining by nearly half. For the conflict as a whole, proponents and opponents were even in terms of arguments on environmental issues but opponents moved to the forefront on these issues during Phase II. Proponents attempted to make their case on economic arguments at the outset, but oddly enough, this line of argumentation fell to a trickle in Phase II when the opposition seemed to steal their thunder with social and political issues.

The analysis also shows that local autonomy, an indicator of the NIMBY syndrome, moved up from sixth in Phase I to second in Phase II of the dispute. This shift was part and parcel of the use of locally oriented arguments by opponents to defeat arguments for the common good made by project promoters. Although arguments for local autonomy were more frequent toward the end, it seems likely that they were motivated by other issues. A closer look at the changing mix of issues suggests that local autonomy was instrumental. The overriding environmental concerns of pollution of the local environment and ecological effects on the forest invigorated the opposition and served to shift public opinion against the WFPP. That these issues were cast in the mold of local autonomy seems to have been a strategic move designed to defeat the powerful interests from outside the community that promoted the project.

5

Subsequent Wood Energy Controversies in the United States

Development in Michigan following the Hersey Controversy

The joint announcement by Consumers Power Company and Wolverine Electric Cooperative that the Hersey project had been shelved created a flurry of interest. The utilities stimulated this response by disclosing that consideration was being given a number of unsolicited offers. More than a dozen communities reportedly tried to fill the vacuum created by the decision to withdraw from Hersey. A spokesman for Consumers cautioned, "... this time we are going into a location where we won't get any surprises."[1] The community picked would likely be the one that most aggressively pursued the project. On 16 June 1980, Consumers Power announced that it had received unsolicited offers from ten communities interested in becoming the site for the demonstrational WFPP.[2]

One such community was Evart, a town not more than fifteen miles from Hersey. The Evart city manager had expressed disappointment with the withdrawal decision and optimism that Evart would be the prime site when the project is revived.[3]

The county commissioners of Ogemaw County endorsed the idea of siting the WFPP in their jurisdiction. However, it was evident in the wording of the endorsement that the issues in the Hersey dispute had not escaped their notice. Acceptance of the plant was to be contingent upon submission of an overall plan by the Michigan DNR and U. S. Forest Service officials showing that wood for the plant "will be cut under good management practices and that such practices allow for reforestation and for supply of wood products for local residents and for any other wood industry within the area."[4]

Opposition began to form in many of these communities, in some cases with the assistance of CRUF. CRUF advertised its willingness to assist communities in scrutinizing the WFPP.[5] It was soon called upon to advise a number of communities being considered as potential sites. For example, a meeting was held at Farwell, a community in adjacent Clare County, after it was announced as a candidate. Citizens Against Chemical Contamination sponsored the meeting, inviting representatives from CRUF, Consumers Power, and Wolverine Electric Cooperative.[6] The utilities did not participate.

The WFPP was debated again at Bellaire in August 1980 in connection with possible siting at Ellsworth in Antrim County. The director of the Michigan DNR, Howard Tanner, was called upon to defend the idea at this meeting. The Grass River Committee had organized out of concern about the effects of the proposed power plant. CRUF also provided advice to this group. Issues debated at the meeting included effects on the forest of large-scale wood harvesting and possible decline in water quality due to erosion from cut-over land. Quality of water is especially important in Antrim County where Ellsworth is located because of its many clear lakes and the tourist trade they provide. Consumers Power disclosed to Ellsworth residents that the list of possible sites would soon be narrowed from twelve to three.[7]

The Ellsworth meeting may well have been the last straw. The fact that citizen opposition groups were forming in sites seriously considered by the utilities must have been viewed with chagrin. Even the director of the Michigan DNR could not quell the uprising. That the utilities considered a number of unsolicited offers after the Hersey experience may have been a face-saving measure. In any event, a month later Consumers announced that it would delay for two to three years any further development of a wood-burning power plant in Michigan. John Selby, board chairman of Consumers Power, indicated the decision was due to the utilities' poor financial condition and the decision by Wolverine Electric Cooperative to withdraw from the project. The Rural Electrification Administration had refused to let Wolverine Electric participate in such a major capital expenditure program. At the expiration of the delay the project would be "re-examined from the standpoint of feasibility, need, cost and other factors."[8]

For the next several years there was a lull in wood-electric power development in Michigan. Memories of the energy crisis began to fade as prices and supplies stabilized. Some facilities were underway, such as the Dow Corning plant in Midland (a topic to be dealt with shortly). But as an issue for state policy and planning, electricity from wood did not generate much enthusiasm. All the same, the wheels of government had ground inexorably toward alternative energy development after the energy shocks of the 1970s and the

Entrance to the Dow Corning SECO plant at Midland, Michigan.

An aerial view of the Dow Corning Corp. SECO plant located at Midland, Michigan. *Courtesy of the Dow Corning Corporation.*

outcome would shortly be felt. Federal tax incentives and regulations requiring that utilities buy power from independent producers soon spawned new proposals. Even though the new development would involve corporations not in the utility business and despite the fact that the proposed facilities would be scaled down to about half that proposed for Hersey, the concept once again met with stiff citizen resistance. The controversy at Indian River described in the next section proved that Hersey is not an isolated case.

One large-scale wood energy project that did succeed without public opposition was the Dow Corning SECO (steam and electric cogeneration) facility. It was sited in a town already very familiar with industrial facilities, including the now defunct Midland nuclear power plant, where the economic benefits of the project would be unencumbered by environmental concerns. Furthermore, the developers made arrangements for a wood supply that would not require public decisions about allocating wood from state forestland. This facility, the first of its kind in the chemical industry, and the institutional plant at Central Michigan University were heralded as examples of job-creating development in job-poor Michigan and as a sensible use of the state's abundant forests.[9] Each is described in turn below.

Recognizing the multiple advantages of producing its own steam and power in an era of energy uncertainties, Dow Corning Corporation began construction of a 22.4-MW wood-fired cogeneration plant in March 1980.[10] The WFPP began producing steam in November 1982, electricity in December 1982, and was officially christened in January 1983. It cost $35 million to build and produces adequate steam and electricity to meet all the needs of Dow Corning's largest manufacturing facility for silicone products. Although primarily intended to save up to 25 percent on the plant's $10 million-a-year energy bill, it had the benefit of adding about seventy jobs to the work force. The basis for the cost savings is the efficiency of cogeneration which utilizes heat energy that is generally rejected as waste heat in a conventional power plant.

The new plant became the largest wood-burning power plant in Michigan's lower peninsula. It reportedly saves five hundred thousand barrels of oil a year. Although it can burn natural gas or oil, the plant is designed to be fueled by wood with a 10 percent assist from natural gas. It was expected to burn one hundred sixty-five thousand tons of wood each year from forests within a seventy-five mile radius of Midland. About half the wood was to be waste from forest products operations. The rest would come from a nearby four thousand acre forest owned by Dow Corning, wood purchased from

Problems with wood chip piles at the Dow Corning SECO plant were solved by construction of storage silos. *Courtesy of the Dow Corning Corporation.*

The completed silos for wood chip storage at the Dow Corning SECO plant.

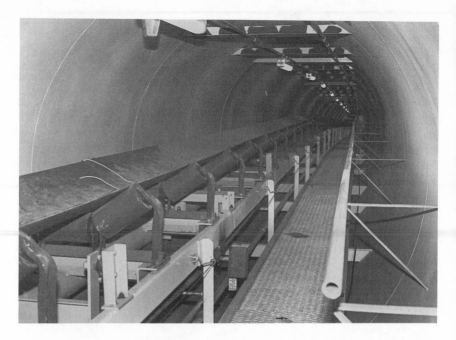

The conveyor belt for transporting chips to the boiler at the Dow Corning SECO plant. *Courtesy of the Dow Corning Corporation.*

The truck dump in action at the Dow Corning SECO plant. *Courtesy of the Dow Corning Corporation.*

private landowners, and the remains of forests in the Mio area that had burned in a massive fire. Professional foresters were hired to direct wood gathering. As of 1988 the plant was consuming three hundred sixty thousand tons of wood chips annually with about 50 percent coming from sources other than forest management.[11]

The Dow SECO plant, as it is called, received widespread recognition. Besides coverage in a number of national magazines and journals, it was cited as one of the ten outstanding engineering achievements is the United States for 1983.[12]

Central Michigan University has established itself as a leader among state institutions in energy conservation. A major reason for this distinction is a $3.5 million state-of-the-art wood burning plant that fires the boiler to heat the school.[13] Construction began in May 1984, the plant coming on-line in June 1985. The cost savings were expected to be dramatic, estimated simple payback being put at three to four years. Fuel requirements were forecast at forty-four thousand tons of wood chips annually, most of it coming from landowners within a fifty mile radius of Mt. Pleasant or from waste wood and wood residue sources in the vicinity.

The university was cautious in designing and presenting the project to the community. Traffic to the plant takes a route that bypasses the highly populated areas of the campus and the town. The ten to fifteen new jobs and the injection of over $1 million per year into the regional economy were strong selling points. A natural resources coordinator was hired to assure that wood is harvested according to good forestry practices.

As in the Dow Corning case, carrying out the project did not require public decisions that might have inspired citizen opposition. The much smaller scale of the project meant that its impact on local natural resources would be of a lower order of magnitude. Finally, Mt. Pleasant is a university town and not a center of tourism. The community welcomed the economic benefits of the facility without apparent worry about effects on local air quality or the forests surrounding the community.[14]

Despite the apparent successes at Midland and Mt. Pleasant, the 1980s must be viewed as a period of stalemate for large-scale wood energy development in Michigan. A number of projects have been proposed largely because of tax incentives and the Public Utilities Regulation and Policy Act (PURPA) requirement that utilities must buy power from independent producers. However, community resistance has continued in some cases, and in others bottlenecks have developed in financing projects. The latter trend is largely due to the financial travails of Consumers Power Company—the utility

Two views of the WFPP at Central Michigan University.

that serves most of the out-state areas where siting WFPPs is feasible. Consumers hovered near bankruptcy in the early 1980s as a result of its tenacious desire to complete the Midland nuclear power plant. Banks were reluctant to loan money to WFPP developers given the uncertain financial picture for Consumers, which would be purchasing their power.[15] The facilities proposed in the mid-1980s have been smaller in size than Hersey or the Dow SECO plant. Most were designed to take advantage of waste from forest products industries in the region in addition to harvesting of forests. The apparent trend in WFPPs is toward smaller cogeneration facilities that will be more efficient energy converters.

A modest movement has also arisen in the business community to drop utility service altogether by producing power within the firm. Energy Michigan, Inc. is a nonprofit association formed in 1984 with a mandate to promote cogeneration in Michigan businesses.[16]

Table 7 summarizes projects in progress since the Indian River controversy described in the next section. Listed are location, size of proposed plant, name of the developer, expected date of completion, and estimated wood requirements.

TABLE 7

Planned Major Wood-fired Power Plants in Michigan, 1983–1986

Site	Size (MW)	Developer(s)	Expected Completion Date	Wood Required (green tons)
Hillman, Montmorency Co.	12.5	Hillman Energy, Inc. Interwest Energy, Inc.	December 1985	200,000
Lincoln Twp., Alcona Co.	15	Viking Energy Corp.	December 1985	180,000
McBain Twp., Missaukee Co.	15	Viking Energy Corp.	December 1985	180,000

Source: Downstate newspaper accounts which, in turn, are based on press releases from developers or the Michigan Public Services Commission.

It is interesting to note that two of the three sites are very sparsely populated and thus less likely to be the locus of community opposition. McBain has become something of a boomtown because of the integrated development of forest products and cogeneration operations at its industrial park. Yet even in this remote, sparsely populated community the old worries about wood supply are

About two-thirds the rated size of the proposed Hersey plant, the McBain WFPP burns wood chips as well as residues from wood processing industries.

The mountain of wood chips at the McBain WFPP.

emerging. A retired forester who manages a mill in McBain stated, in an otherwise glowing account of economic revitalization for the area, that availability of timber is becoming a problem. "We're out 108 miles now. We're talking about getting wood from the Upper Peninsula."[17] Wood chips from the operation, amounting to about seven hundred tons per day, were used to make particle board and to fuel the Dow Corning SECO plant.

As of October 1988 none of these projects had been completed. The WFPPs at Lincoln and McBain were under construction and expected to be operational in the spring of 1989. The Hillman plant was near completion and would begin producing power in January 1989. Its current developer was Stockmar Corporation. The scale of the WFPP increased since the planning stage, subsequently being rated at 17.5 MW versus the proposed 12.5 MW.[18]

The McBain WFPP was completed in time to provide some excellent pictures for purposes of this book. Rated at 16 MW, it opened early in 1989. The project cost about $25 million to complete and is held by a limited partnership formed by Energy Factors, Inc. of San Diego, California and CRS Sirrine of Houston, Texas. The McBain WFPP sells electricity to Consumers Power Co. under a 32-year agreement. Twenty-one people are directly employed at the plant and $2 million will annually be spent in the regional economy to purchase fuelwood.[19] The accompanying photographs clearly show what this technology looks like. Although somewhat smaller than the proposed Hersey WFPP (see p. 100), the design of the McBain plant is quite similar.

That the planned WFPPs were still not on-line in 1988 cannot be laid at the doorstep of government. Incentives provided by the federal government and programs administered by Michigan have been instrumental in encouraging private interest. As just mentioned, projects have not gotten off the ground largely because of controversy or lack of confidence by financial institutions. Both the Michigan Public Service Commission (PSC) and the Energy Administration of the Michigan Department of Commerce (an office recently merged with the Michigan Public Services Commission) have been unwavering in their commitment to wood-electric power development. The Michigan Biomass Energy Program, for example, was established in the Energy Administration in January 1984. Its purposes were to promote the adoption of wood energy in commercial, industrial, and public facilities where economically feasible and environmentally acceptable, and to identify and resolve problems related to widespread wood energy use. In order to implement PURPA, the Public Service Commission has been

The 16-MW McBain WFPP finally opened in 1989.

involved in monitoring and facilitating WFPP projects. Yet the bottlenecks alluded to are not the only ones currently facing wood-electric power development. Any future projects must deal with the legacy of widely reported problems with the world's largest wood-burning plant, a 50-MW plant at Burlington, Vermont—the topic of the final section in this chapter.[20]

The Indian River, Michigan WFPP Controversy

The proposal to site a WFPP in the northeastern Michigan community of Indian River touched off a controversy similar to the Hersey case.[21] The plant was to be half the size of the Hersey WFPP and would have relied on wood gasification to produce energy. Nevertheless, the proposal was defeated by an entrenched citizen opposition. The many parallels to the Hersey case tempt one to believe that ironclad laws of social behavior are at work. At the very least, the similarities suggest a general social pattern developers can often expect to encounter in response to WFPP siting proposals. Here follows a brief account of the Indian River controversy.

The truck dumper and the large pile of wood chips behind the McBain WFPP.

Another view of the truck dumper showing the wood chip conveyor at the McBain WFPP.

Spurred in part by federal tax incentives for alternative energy development (the Public Utilities Regulation and Policy Act of 1978, also known as PURPA), a group of businessmen formed in 1983 under the name Primary Power. The company sought to take advantage of PURPA's requirement that utilities must buy electricity from small power producers. The proposed site for the 12-MW wood gasification plant was two miles north of Indian River, a town of about two thousand located in northeastern Michigan's Cheboygan County (see figure 2). Unlike direct burning in the Hersey plant design, gasification utilizes the gas from heating wood-chips and sawdust. The gas is directed into a steam boiler which, in turn, drives the turbines for an electrical generator. The plant would have used an estimated two hundred thousand gallons of water and three hundred tons of wood by-products daily. Promoters touted the addition of seventy full time jobs to the local economy, the use of local contractors during construction, and the advantages to the tax base in an area that relies primarily on spring and summer tourism. In addition, the plant promised to be a boon to about twenty area sawmills by creating a market for their surplus sawdust and wood chips.

Unfortunately for Primary Power, Indian River is located near the southern end of Burt and Mullett Lakes—large lakes that are heavily used for recreation and tourism. Just as in the Hersey controversy, in-migrants, buoyed by a tenacious concern for environmental quality, proved to be a force in local politics. Soon after the power plant was proposed, the Committee to Save Our Resources and Environment (SORE) formed to fight the project. Its leader was an in-migrant to the community from Detroit who was able to mobilize local opposition to defend local environmental values. The group conducted a mail campaign in an attempt to influence other residents to oppose the plant. They sought, as had the opponents of the Hersey plant, to effect their strategy at the local level.

SORE supporters considered the power plant to be an impending environmental disaster, notwithstanding assurances from the DNR. Air pollution was their foremost concern at the outset. The list of uncertainties in WFPP development had expanded since the Hersey controversy to include pollution from burning wood. They expressed the familiar view that the sole purpose of the project was to line the pockets of Primary Power investors with federal tax breaks. SORE gained support from people living outside the community who owned cottages on Burt and Mullett Lakes. Some of them had a substantial financial interest in the area. For example, one of the colitigants with SORE was a real estate developer who feared the project's impact on a condominium complex and a new eighteen-hole golf course.

SORE looked for any forum that would provide the opportunity to take issue with the proposal just as the Hersey opponents had done. When administrative measures failed, the opposition used litigation to achieve its aims—another parallel with the Hersey case. Primary Power had requested a waiver from the Michigan Air Pollution Control Commission to proceed with construction. The ensuing review by the Michigan Environmental Review Board and concomitant public hearing would provide opportunities for SORE to enter the technical debate. In the meantime, the Cheboygan County Zoning Commission acceded to a request by Primary Power for a zoning permit. As a result, SORE filed suit against both of them in December 1983. A Cheboygan County circuit judge issued a summary judgment in November 1984 voiding the special use permit of the Zoning Commission. The judge found that the commission had not followed established standards and procedures for approving a special use permit. Although the WFPP had been defeated on procedural grounds, SORE regarded it as a victory for the environment. The attorney for the group was poised to raise environmental arguments if the procedural tactic failed.

By the time of this judicial decision SORE had adopted the guiding light of the Hersey debate—the availability of wood. In the wake of its victory SORE leaders moved beyond the issue of air pollution, focusing on the effect that the plant and others that might follow it would have on the forests in northern Michigan. They suggested that reforestation projects would not adequately replace the timber cut for the power plant.

In addition to the newcomer-activist parallel with the Hersey controversy, the controversy at Indian River demonstrates the tenacity of local autonomy as an issue. Opponents seemed to be motivated by the thought an outside corporation was foisting the project on the community as much as by concern for environmental degradation. When questioned about the claim that the plant would meet state air pollution standards, a spokesman for SORE said it may meet state requirements but it doesn't meet Indian River standards. Arguments about the value of the contribution of power produced by the plant were also debated in terms of local autonomy. The leader of SORE simply pointed out that Indian River doesn't need the power because it has plenty of electricity available from Consumer Power Company. Nevertheless, as in the Hersey case, leaders of the opposition showed a cosmopolitan orientation too,

The Burney Mountain WFPP is a clone of the Mt. Lassen WFPP at Westwood, California. They were developed simultaneously and both opened in October 1984. *Courtesy of Pacific Energy.*

arguing that damage to the forests of Michigan would be travesty for all.

The California Controversies

Controversies in the United States have attended the building of WFPPs in the West as well as the Midwest. This is somewhat curious because the western United States has longstanding experience with wood-burning facilities in the forest products industry. Once again it seems that the scale of the proposed power plants and their potential threat to other values precipitated the conflicts.

California is recognized as a national leader in alternative energy development. Witness the windmill farm in Altamont Pass, for example, in the breezy coastal hills east of San Francisco. By 1985 several WFPPs had been built and many were in various stages of planning and construction. Ultrapower, Inc.—a subsidiary of Ultrasystems of Irvine, California—became the state's leading entrepreneur of WFPPs. Its first two plants were sited at Burney and Westwood. In 1986 another was built at Blue Lake. The Burney and Westwood plants rely on logged slash for about 40 percent of their

The Mt. Lassen WFPP was built at Westwood despite modest community opposition. Rated at 11.4 MW, it opened in October 1984. *Courtesy of Pacific Energy.*

fuel, the rest coming from mill wastes and timber logged for fuel. The Blue Lake plant primarily burns redwood bark from local sawmills. All three are 11.4-MW cogeneration WFPPs. Chinese Station, a 25-MW WFPP, has also been built by Ultrapower near Sonora, California. Ultrapower sells the power it produces to Pacific Gas and Electric.[22]

This upsurge in biomass energy development has prompted the California Energy Commission to produce and update a map of planned and existing biomass energy facilities in the state. Although development was brisk in the 1980s, it has not been trouble free. Both the Burney and Westwood plants were plagued with unexpected costs. Most of Ultrapower's $4.4 million losses in the quarter ending 31 January 1984 were due to start-up problems including major repairs to the air pollution control systems of these power plants.

WFPPs have not been a major issue for environmentalists. In 1985 the Sierra Club had no national policy on wood-burning electric power plants, nor had it received much comment on the issue. Although statewide and national organizations may have ignored the environmental aspects of WFPP development, this has not been the case at the local level. The defeat of one such plant will shortly be described. Neighbors of the power plants that were built generally did not object prior to construction, but later began to express worry about the potential for air pollution and lack of job creation. When Ultrapower's plant at Westwood broke down in December 1984, "spewing black ash and cinders across the snowy landscape," a petition was immediately circulated by outraged residents demanding closure of the plant until its problems could be solved.[23]

A study of controversies in two contrasting communities in northern California reveals many of the same trends as observed at Hersey and Indian River including the threat to local autonomy and the importation of environmental values via the population migration turnaround. The following discussion is largely a summary of this study.[24]

California's response to the energy crisis was similar to that of other states. Many, especially those in the rural forested areas of the state, adopted the wood stove for residential space heating to deal with rapidly rising prices for fossil fuels. Government programs to encourage large-scale wood energy development had their effect here as well. Motivated by federal tax incentives, Ultrapower, Inc. proposed to build two 11.4-MW wood-fired power plants in northeastern California in 1981 (see figure 3). Westwood, in Lassen County, was soon selected as a site, followed nine months later by Quincy in Plumas County. Ground breaking took place at Westwood at about the time of Quincy's selection. However, the outcome would be different at Quincy. The controversy there lasted about ten months. It began with the first permit application by Ultrapower and ended in 1983 when the developer withdrew the project.

At the outset Ultrapower touted the economic benefits of the facility in both communities. These were said to include fifty to seventy-five new jobs, an increase in the local tax base from property and salaries, the economic impact of $20 million invested for construction, and improved forest management. At the time California was suffering from a depression in the timber industry. No doubt Ultrapower expected the WFPP's economic benefits would be a strong inducement indeed for the selected communities. As it

Figure 3. Location of Quincy and Westwood, California

happens, relative poverty in these areas had already forced many residents to depend on wood stoves for heating. The study humorously notes that "the unifying architectural theme in Westwood and Quincy is firewood in multiple-cord stacks."[25]

Not surprisingly, the wood appetite of these power plants was viewed by some as a threat to their fuel supply. This perception proved to be the crucial issue in the successful Quincy protest. The previous protest of a fee imposed on woodcutting by the Forest Service underscored local concern for the local wood supply. In a fashion reminiscent of the Hersey dispute, Quincy activists broadened plant opposition to nearby communities, making fuelwood depletion a county-wide issue.

To a lesser extent local autonomy was an ingredient of the dispute. Many were put off by the "slick big-city image that Ultrapower projected."[26] Plant promoters were viewed as southern California outsiders. As in the Indian River controversy, the plant was

A wood chipping operation in a California pine forest. *Courtesy of Pacific Energy.*

A truckload of chips is dumped at the Mt. Lassen WFPP in Westwood, California. *Courtesy of Pacific Energy.*

portrayed as a attempt by a greedy corporation to take advantage of federal tax incentives. There was also some animosity toward Forest Service policies and officials from outside the community, but loss of control over the local firewood supply was seen as the most significant threat to local autonomy.

A number of environmental issues crept into the controversy. There was some concern about the prospect of air pollution in times of the thermal inversions common to the area, and of water pollution due to waste heat from the power plant. The explicit worry that the plant might eventually burn coal and be much more polluting presents a parallel with the controversial possibility at Hersey that the WFPP might burn municipal refuse. Groundwater depletion was cited as a potential problem. The effects of wood removal on forest nutrient cycling was also recognized and debated.

Some of these issues, such as depletion of firewood and air pollution, were raised in the Westwood dispute, but because of social differences here as compared with Quincy, attempts to organize against the power plant failed. In both cases a small citizen group formed, but only one succeeded.

On the basis of two cases, the authors of this study suggested some reasons why the WFPP siting at Westwood succeeded while the one at Quincy failed. Quincy is larger, more differentiated, wealthier, and better organized.[27] It harbors a diverse collection of community organizations and associations totaling over one hundred and fifty, including some environmental groups. Its population was 4,451 in 1980, over twice that of Westwood's 2,117 residents. Quincy has a community college and serves as the county seat. Road access to the town is good. Westwood does not compare well in any of these areas. It is much less accessible and has proportionately fewer organizations and associations. Poverty is more evident in Westwood with nearly 19 percent of its residents having incomes below the poverty line in 1979.

Since 1975 a higher percentage of residents had moved into Westwood (47 percent) than had moved into Quincy (40 percent). However, three times as many Quincy residents (about one in seven) had moved to the town from outside the state. This kind of diversity and the values it implies helped fuel the protest.

An environmental group in Quincy had previous experience with natural resource conflicts. Several members of the group formed a committee to scrutinize the proposed WFPP. This committee soon became the lightning rod of opposition. As in the Hersey case, these activists applied their know-how in research, organization, and intervention to opposing Ultrapower in public proceedings. The

The Mt. Lassen WFPP at Westwood, California, in winter. *Courtesy of Pacific Energy.*

experience and knowledge of a few people was enhanced by access to other organizational networks in the community. Westwood, on the other hand, had no effective group involved in environmental issues, a fact that enabled the plant to be pushed through without knowledgeable objection.

The authors of the comparative study of Westwood and Quincy concluded that three major factors are responsible for the divergent responses to the project—variations in economic well-being, level of social organization, and relative centrality.[28] Quincy had relative advantages in all three. It was wealthier, better organized, and experienced in social protest, and it was a center for decision making that would affect the outcome of the proposal.

Finally, for purposes of comparative study, it is well worth noting parallels on the matters of expertise as a political resource, the population migration turnaround, and the importance of local decision making.

Quincy plant opponents hired a University of California at Davis student as an environmental consultant. This outside environmental expert apparently helped with the research effort that enabled the citizens committee to effectively participate in technical debate about the WFPP. On the other side of the ledger, residents of the community learned that an important company consultant had been employed at very senior levels of the Forest Service. The perception that an old-boy network was involved may have been a source of distrust of the Forest Service.

As in the Hersey case, newcomers joined forces with old-timers. Together they sought to preserve shared values against the interests of developers from outside the community. The essential ingredient, however, was the newcomers' organizational skills and connections to outside networks. Using the language of the authors of the California study, "the [Quincy] reverse migrants brought to a rural area sophisticated urban organizing skills to be used against a sophisticated urban company."[29]

The unrelenting activity of opponents at the local level was decisive. County supervisors and the planning commission proved vulnerable to the political persuasion of the plant opponents. Local politics was the critical arena for decisions on air and water pollution. As was true of the state Natural Resources Commission at Hersey, the U.S. Forest Service failed to respond to the concerns of the citizen opposition group at Quincy.

The essential finding of the California study is well worth quoting here to conclude this summary. It provides some general insights into the process of citizen opposition in nonmetropolitan communities and stands as both a description and a prescription.

As the population of resource-dependent areas has changed, it has become increasingly clear that people who live adjacent to resource areas also have deep concerns over their management. These concerns are not new, but the realization that distant bureaucracies and local governmental elites can be forced into action by public pressure is relatively new.[30]

A COMPARISON OF THE MICHIGAN AND CALIFORNIA CASES

A further examination of the California study reveals some differences with the present study that merit consideration. The California researchers also found local protest to be an essential element of the strategy employed by opponents of WFPPs. Their claim that rural environmental protest is a new strategy is subject to interpretation.[31] Rural environmental protest predates the first earth day—for example, the controversy over the federal proposal to site a high-level radioactive waste disposal facility in bedded salt formations near Lyons, Kansas, in the late 1960s. They seek to establish the idea that conflicts over local natural resources can only be understood in terms of "the existence of nonstatutory claims to property." WFPP controversies contain two elements which contradict this assertion. First, public lands are usually involved. Since all citizens are technically owners of such lands, they clearly have a vested interest in their proper management. Second, the environmental problems created by such large-scale developments generally do not recognize and certainly do not stay within statutory boundaries. If air pollution from a WFPP is carried within the airspace of a property owner, it would seem to be a matter of statutory ownership. The essential problem is not ownership of land but the quality of the environment, a concern that transcends the issue of who owns what and therefore who has a right to make decisions about the property in question.

On the basis of two cases the authors of the California study would abandon sociological research beyond an assessment of relative rights. They state, for example, that: "communities mobilize partly to protect these nonstatutory claims to property . . . instead of searching for an ideological basis to explain community protest attention might be better directed toward identifying disputed claims to property."[32] The ensuing comparision with the Michigan cases suggests that thus limiting the study of natural resource conflict may be a serious mistake.

Notions of ownership and rights are particularly important in the western United States where water is a paramount concern. Applying this concept to natural resource conflicts of all kinds imposes a western bias that very well may misconstrue controversies in general. Certainly in the Michigan cases, ownership or presumed property rights were not issues. Local autonomy (or NIMBY) was important as a strategy. However, the debate revolved around the environmental values at stake and seldom was couched in terms of rights.

The California researchers also seek to dismiss the values of newcomers, claiming instead that the organizing skills and connections to outside networks possessed by in-migrants were the essential factor.[33] This is a curious argument. Without a perception of value and the motivation to act to conserve the object which is valued, it is difficult to imagine how newcomers might possess an ideology that would prompt them to organize a protest. Denying the fundamental importance of values is tantamount to throwing out the baby with the bath water. I would argue that such conflicts demonstrate the crucial importance of values—not just community structure—in community mobilization.[34]

Both the Michigan and California studies contain cases of conflict that succeeded and failed. In California the WFPP proposed for Westwood was built while the one to be sited at Quincy was defeated. Although plants were successfully opposed at Hersey and Indian River, Michigan, WFPPs were built without controversy at McBain and Midland, Michigan. Data for communities where a WFPP was defeated are arrayed in table 8 and for those where one was built in table 9. These data are considered below in relation to the arguments raised in both studies.

Some seven years have elapsed since the culmination of the California controversies at Westwood and Quincy. With the actual controversies receding into the past and with this period of time to consider its causes and effects, the authors of this study conclude their published account with an explanation based on three factors—economic well-being, social organization, and centrality.[35] Measures of these factors for which they marshall data include, respectively, poverty rate, number of business and social organizations, and access to county-level government. Quincy had relative advantages in all three, hence the reasons for the success of its protest against the WFPP according to these researchers. A cursory look at the data in tables 8 and 9 demonstrates how preliminary and perhaps ill-conceived this explanation may be.

Economic well-being. To use median income as an indicator, economic well-being was relatively low in McBain ($10,417), just as it was in Westwood ($12,628). Perhaps the allure of jobs related to the WFPP was overwhelming for both communities. On the other

TABLE 8

**Demographic, Social, and Economic Characteristics of Sites
where Wood-fired Power Plants Were Defeated:
Quincy, CA; Hersey, MI; Indian River, MI**

Variable	Quincy, CA	Hersey, MI	Indian River, MI (Tuscarora Twp.)
1980 population	4451	1228	1952
Income Characteristics, 1979			
Median income	$18,255	$14,957	$14,886
Persons below povery level	6.2	13.4	10.9
Households receiving public assistance	11.0	13.0	7.6
Households with social security income	24.9	35.1	42.2
Employment and Education			
Males over 16 who worked in 1979	61.7	67.4	66.7
Females over 16 who worked in 1979	50.7	47.1	49.3
Persons over 25 with some college	40.7	16.8	26.5
Migration			
Persons living out of county in 1975	39.7	36.1	37.4
Persons living out of state in 1975	14.9	3.7	3.4

Figures for all variables except 1980 population and medium income are expressed as percentages.

Source: U.S. Bureau of Census, Population and Housing Summary Tape 3A for these communities in California and Michigan.

hand, Midland ($20,142) had the highest economic well being of all the communities studied. Yet it became the uncontested site of a relatively large WFPP (about twice the size of the plants proposed for Westwood and Quincy). Midland is the home of the Dow Chemical Company and the Dow Corning Corporation (developer of the WFPP). Given longstanding experience with large industrial facilities, a WFPP must have seemed innocuous to its citizens. A highly controversial nuclear power plant, one that nearly bankrupted Consumers Power Company, lies uncompleted at the edge of town. A WFPP would seem quite tame next to the controversy engendered by the nuclear power plant. Be that as it may, the comparative wealth of Midland and its past connections with large industry would appear to foretell acceptance of a WFPP.

The explanation that economic well-being is a cause of local natural resource protest is therefore unpersuasive, at least when median income is considered. Sites with high income built WFPPs. Others with low income defeated them. But how about percent of persons below poverty level as an indicator of economic well-being?

TABLE 9

**Demographic, Social, and Economic Characteristics of Sites
where Wood-fired Power Plants Were Built:
Westwood, CA; McBain, MI; Midland, MI**

Variable	Westwood, CA	McBain, MI	Midland, MI
1980 population	2117	530	4031
Income Characterstics, 1979			
Median income	$12,628	$10,417	$20,142
Persons below povery level	18.8	11.7	12.7
Households receiving public assistance	11.1	5.3	10.5
Households with social security income	27.4	45.0	25.9
Employment and Education			
Males over 16 who worked in 1979	73.7	71.8	80.5
Females over 16 who worked in 1979	50.7	40.6	48.3
Persons over 25 with some college	33.2	15.5	34.1
Migration			
Persons living out of county in 1975	47.1	16.6	29.8
Persons living out of state in 1975	3.5	0	9.6

Figures for all variables except 1980 population and medium income are expressed as percentages.

Source: U.S. Bureau of Census, Population and Housing Summary Tape 3A for these communities in California and Michigan.

Percent of persons living below the poverty level was relatively low in Quincy (6.2 percent), suggesting that high levels of economic well-being portend a successful protest. However, Midland at 12.7 percent has a poverty rate more comparable to Hersey (13.4 percent) and Indian River (10.9 percent) where plants were successfully opposed. McBain's poverty rate (11.7 percent) was considerably lower than Westwood's (18.8 percent); however, it did not evidence even the meager attempts at protest by some Westwood citizens. Among all the communities studied, Hersey had the highest poverty rate at 13.4 percent, after Westwood's 18.8 percent, yet it waged a protracted and successful bid to keep the WFPP out.

Percent of households receiving public assistance is another possible indicator of economic well-being. Among all sites, McBain had the lowest figure (5.3 percent) and Hersey the highest (13 percent). At first blush this would suggest the opposite relationship—low levels of economic well-being lead to successful rural protest. However, once again there is a wide range of variability. Indian River was the next lowest (7.6 percent) but rebuffed a WFPP. All of the sites where a WFPP was built were at or below the figure for Quincy which rejected the proposed plant. So we are no better off in postulating this measure of economic well-being as an explanation.

One final measure of economic well-being provides some interesting results. Figures for percent of households receiving social security show the contrasting cases of McBain and Indian River with the highest figures, respectively 45 percent and 42.2 percent. Each site had a relatively aged population. In the case of Indian River it was probably retirees that had permanently moved to cottages on adjacent lakes or into town. McBain's elderly are very likely the result of earlier out-migration that left the area with a larger proportion of its population in the older age groups. The lack of correspondence between this variable and natural resource protest is evident in the mixed figures for this indicator—Quincy (24.9 percent) and Hersey (35.1 percent) are not much different from two sities where plants were built, Westwood (27.4 percent) and Midland (25.9 percent).

The highly mixed results of these various measures show how dangerous it is to make an explanation based on a comparison of just two contrasting cases. Obviously, by extending these indicators to the Michigan cases, we are left to conclude that level of economic well-being by itself has limited potential for explaining natural resource conflict. Economic well-being would seem to be neither necessary nor sufficient for such an explanation.

The present study of wood energy development in the United States points to a different explanation. Lack of economic well-being is not necessarily reducible to convenient aggregate measures. Other, noneconomic values are very important to people in forested regions. These values clearly affect their behavior relative to such decisions. The statement from an opponent of the Hersey WFPP quoted on p. 197 suggests that poverty by itself is a bankrupt indicator of natural resource conflict (excuse the pun) because it fails to take into account the noneconomic values of rural residents. Note that "right" in this statement refers to political rights and not to property rights. The term is clearly referenced to involuntary risks to quality of life in the community.

Social organization. As a indicator of social organization the California researchers counted businesses and community organizations. They then concluded that because of the abundance of organizations in Quincy and the lack of same in Westwood, this measure obviously provides an explanation for the relative success of their protests. Once again the Michigan cases confound this explanation, for they reveal great variability and suggest that we must look deeper for the cause.

Hersey and Indian River are small towns with few businesses or community organizations but, like Quincy, they defeated WFPPs. McBain was similar to Westwood in lacking social organization and accepting a WFPP. However, Midland, with an abundance of businesses and community organizations, accepted a WFPP. Perhaps some of the WFPP promoters made the same mistake as the authors of the California study. They may have taken a cusory look at Hersey and Indian River, concluding either that these communities would welcome the economic benefits of the project or would fail to mount a protest. Having been to Hersey a number of times, I can attest that their is nothing about this community suggesting a penchant for protest. It possesses a mere handful of storefronts and no indication of the Lions, Elks, and Rotary Club. Nearby Evart and Reed City provide such organizations, but Hersey is a sleepy little town that has hardly changed in the last decade (see p. 104).

The idea that some critical number of organizations makes protest either possible or likely does not fit with the facts. I would again argue that an explanation must be sought in terms of community values. Large-scale projects can succeed in large, highly organized communities and fail in remote, rural communities having little apparent social organization. The determinant of whether a protest will arise depends on how the facility is perceived to affect values held dear by members of the community, oldtimers and newcomers alike. The Michigan cases show that if local values are threatened even the smallest community with very few businesses or organizations can overwhelm developers and state agencies. Such communities may take advantage of networks or otherwise seek help from outsiders but, in any event, they seem able to mobilize the political resources for an effective opposition.

One additional factor relevant to the ability to organize is education. Tables 8 and 9 show the percentage of the population at each site that had some college education. Quincy had the best-educated population with 40.7 percent having attended college. At Midland, a town of comparable size with a well-educated workforce in its large industries, the figure is 34.1 percent. The fact that a WFPP was built at one but not the other demonstrates that education by itself is an inadequate explanation. Indeed, the figures for Hersey (16.8 percent) and Indian River (26.5 percent) are even lower than Westwood (33.2 percent) where a WFPP was built. McBain at 15.5 percent is the lowest but not significantly less than Hersey. It would seem that having some members of the community who are well educated is a necessary but insufficient condition for a successful protest.

Education must be reckoned in terms of its effects on values and ideology. It is entirely possible to be well educated and lack a proper sense of values. The scandals on Wall Street and among savings and loan institutions in the United States have aptly demonstrated this fact. The population migration turnaround has brought urban defectors to rural areas. Some were well educated but they also possessed values and ideologies inconsistent with corporate development for profit by the few at the expense of the environmental values that had attracted these migrants to a community. Some of the in-migrants became local experts in the administrative forums where the issues were debated. The high figures for in-migration to Hersey (36.1 percent) and Indian River (37.4 percent) since 1975 are quite comparable to Quincy (39.7 percent). Among this migration stream were some urban defectors who organized the opposition. Midland has a somewhat lower rate of 29.8 percent; however, many of these inmigrants must have been connected with Dow Chemical since 9.6 percent came from out of state. Midland's inmigrants obviously were not moving there for fresh air to breath. The low figure of 16.6 percent for McBain, along with its very small population, accords with the fact that no protest developed there. The only anomaly is Westwood, with the highest figure of all for inmigration at 47.1 percent. The population migration turnaround appears to be a necessary but not a sufficient ingredient for natural resource protest in rural areas.

Centrality. The explanation from centrality was undoubtedly the easiest one for the California researchers to make. After all, Westwood is nestled in the mountains—away from relevant centers of government. It is often snowbound in the winter. Quincy provides instant access to county government. Accordingly, it was easy to adopt this as an explanation for the relative success of their protests. Once again, the Michigan cases confound this convenient explanation.

Unlike Quincy, Hersey and Indian River are more remote from county government; yet they successfully opposed WFPPs. Hersey is about ten miles from Reed City, the Osceola County seat, whereas Indian River is more than twenty miles from Cheboygan, the county seat of Cheboygan County. On the other hand, McBain is ten miles directly south of Lake City, the county seat of Missaukee County. Notwithstanding this good access, it accepted a WFPP without protest. Midland is a center for county government, yet it embraced a WFPP, by contrast with Quincy.

County government is only one forum where crucial decisions are discussed and made. It is a rather unfortunate choice for an indicator because the most effective arena for citizen action is local government. Township zoning boards may have far more to do with a successful protest. Local government was critical in the Hersey controversy. State and county governments lent a deaf ear to the concerns of the Hersey opposition. Judicial forums are also important avenues for protest. The Indian River WFPP was defeated by litigation. Litigation was important in the Hersey case as well.

The fact that the Quincy WFPP met its death at the hands of the county does not stand as an endorsement of county government as a nexus for protest. The Michigan cases suggest that the most important aspect of centrality is local government. Since even the smallest community possesses this factor to some degree, centrality as an explanation is apparently a moot point.

This review of two comparative studies underlines the notion that great care must be exercised when advancing explanations based upon case studies. The California study is valuable for detailing another case of natural resource conflict. However, it is flawed in summarily imposing a conceputual framework (statutory rights) and an explanation of causation (social structure), in the process ignoring findings from other highly relevant studies.

The Michigan WFPP cases reveal the wide variability in social structure that were so readily taken for explanations in the California study. Short of claiming that the present study of large-scale wood energy development has reached the correct explanation, it is fair to say that understanding natural resource conflict lies beyond the pale of imprecise notions and measures of economic well-being, social organization, and centrality. An explanation couched in values and ideology finds support in both of the studies reviewed here.[36] Whether environmental values and their embodiment in ideology are the ultimate explanation remains to be seen. The role of the population migration turnaround as a vector for change in these terms is undeniable.[37]

The Burlington, Vermont 50-MW WFPP

The city of Burlington has been on the vanguard in development of wood for electric power since shortly after the emergence of the energy problem. In 1977 it began by experimenting with a mixture of 75 percent wood chips and 25 percent oil in a 10-MW power plant, reducing cost per kilowatt hour of electricity from three cents to two. Encouraged by this experience, the utility proposed a 50-MW plant to meet the future electrical demand of the city. In 1978

Burlington area voters overwhelmingly approved (by a margin of 70 percent) a $40 million bond issue to finance the project.[38] The total cost was projected at $80 million for the WFPP. On 15 November 1979, just as the Hersey controversy was coming to full throttle, the *Osceola County Herald* carried a front page article bearing the title "Vermont Wood Chip Plant Shelved." Officials of the Burlington Electric Department had instead decided to buy a 25-MW block of power from Ontario Hydro. It is very likely that this report received front page coverage in only two newspapers—the *Burlington Free Press* and the *Osceola County Herald*. The article in the *Herald* also reported that a petition protesting the proposed Burlington WFPP had been submitted to the Vermont State Energy and Public Service Board. An organizer in the opposition at Burlington was quoted as saying "A lot of people are concerned about the effects of this."[39] The temporary outcome must have had an impact on the Hersey controversy, whose participants were made aware that the proposed development at Burlington was also controversial.

Burlington, Vermont, is the state's largest city. It had a population of about thirty-eight thousand when the city began experimenting with wood-electric power in 1977. Because 76 percent of Vermont's land area is forest, wood energy was a logical alternative to consider in the face of expensive imported oil and rapidly escalating costs for nuclear power plants. Moreover, much of the stock of timber was low-grade and seemingly best suited for fuel in a power plant. Burlington Electric Department (BED), a municipally-owned utility, was so pleased with its retrofit of a coal-fired boiler to burn wood and the subsequent cost savings that it aggressively pursued WFPP development. BED proposed to build the largest wood-fueled power plant in the United States, indeed, in the world. The successful bond referendum just mentioned indicated broad-based community support. An unusually good offer from Ontario Hydro and a modest protest, in the form of a petition by seventy residents to the Vermont Energy Board and Vermont Public Service Board, prompted BED to temporarily delay the project. The project would soon move ahead to completion.

It is worth digressing here to describe the initial opposition to the WFPP. Burlington Electric Department had modified its plans for the facility right from the start in order to accede to public concern.[40] For example, the plant's design was changed to allow rail delivery of wood chips because of public disenchantment with the amount of truck traffic the plant would require. This proposal was altered again because of protests over the noise rail cars would make in the area originally designated. Just as the Ontario Hydro deal surfaced, opposition by a citizen group, Vermonters Against the Burlington Plant, culminated in the petition to state agencies. Seventy residents of Lamoille County signed the petition. The group expressed concern that the power plant would place an intolerable burden on the state's forests. Damage was anticipated from the heavy machinery needed to harvest wood. The group also suggested that environmental damage would occur to forestland not accessible because of a lack of roads. Their foremost interest seemed to be the effect on the fuel wood market for home heating. The group felt that residential wood users would be pressured by an inevitable rise in wood prices and decline in availability. They also noted in the petition that the real cost of the plant would be to residents of locations in the state outside the city of Burlington who would be effectively feeding their forests to the city just to stabilize its electric rates. In a fashion similar to the opposition at Hersey, the group cited cosmopolitan as well as local arguments to make its case. Heavy emphasis was placed on the issue of local autonomy (that is, communities outside Burlington being forced to pay the social costs for its electricity). The petitioners called for a statewide referendum on the project based on the assumption that most of the wood required would be harvested from public as well as private forestland throughout the state.

The decision to shelve the plant occurred almost simultaneously with this protest. The superintendent of the BED defended the project despite the decision to defer it, pointing out that the WFPP remained technologically and economically feasible over the long run. By this time, two years and $1.7 million for studies and land purchases had gone into the project. He suggested that the delay might be as long as five years. At the same time the mayor of Burlington affirmed the city's commitment to the concept, suggesting that the eyes of the nation were on Burlington for its pioneering efforts in wood energy. An environmentalist interviewed about the decision to purchase power instead of building the plant termed the deferral a prudent decision.

The public support the idea had received and the various interests seeking to make it a reality brought the planned WFPP off the shelf in a relative hurry. In 1981 the citizens of Burlington again endorsed the plant, voting for a $65 million bond issue to finance its construction. The policy adopted by the referendum involved a strategy of diversity in future supplies, including one quarter each from wood, hydroelectric, nuclear, and coal and other sources.[41] The project received the approval of the Vermont Public Service Board the same year.

Three views of the Joseph C. MacNeil Station, the largest plant of its kind in the world. It is operated by the Burlington Electric Department for the City of Burlington in Vermont. Note the huge pile of wood chips that provides fuel for this 50-MW power plant [in middle photo]. The scale of the plant is apparent by comparison with the automobiles [in bottom photo].

The plant would be sited near the boundary of the neighboring city of Winooski. While building the power plant, the utility sought to assuage public concerns by installing its own rail spur to limit truck traffic and $7 million in pollution control equipment to catch ash.[42]

The Burlington McNeil generating station officially opened in March 1984. Cost of construction amounted to $66 million, less than the $80 million original estimate. First estimates were based on bids from a severely depressed construction industry in the late 1970s. Good project management contributed to the substantial savings. Consumption of wood was estimated at a maximum of fifty hundred thousand green tons per year. An estimated $10 million would be spent annually on wood in the region. This was expected to have a multiplier effect, significantly contributing to the rural economy of northern Vermont. In response to the demand for wood chips created by the plant, the number of whole tree chip operators in Vermont increased from eleven in 1983 to twenty-five in 1984. A new railhead and collection point for wood chips was constructed at Swanton, Vermont, about forty miles north of Burlington. This served to enlarge the area from which timber could be economically harvested for the WFPP and permitted implementation of mandated rail deliveries of wood.[43]

Public officials were careful to respond to citizen complaints about the possible effects of the plant on the forest. BED had also recommended forest harvesting standards. The Public Service Board ruled that BED must have professional foresters to supervise timber harvesting. All whole tree chip harvesters must be licensed by the Vermont Department of Forests, Parks, and Recreation, the agency also charged with monitoring their operations. These regulations were apparently unique to harvesting wood for energy, not previously having been applied to any other type of timber harvesting operation. According to one commentary, "for the first time a utility has been forced by public regulation to consider the environmental and social consequences of resource extraction."[44]

Officials also dealt with vehement objections to the possibility of excessive truck traffic through Winooski. The Public Service Board limited truck deliveries to 25 percent of the fuel required by the plant. In response to this requirement, BED chose to ship the remainder of the wood by rail from Swanton.

Mechanical collectors and electrostatic precipitators were installed in the plant to decrease emissions below standards set by federal and state governments. The McNeil Station power plant was reported to annually emit the following pollutants (all within these standards for power plants):

- 25 tons of particulates
- 98 tons of sulfur dioxide
- 843 tons of carbon monoxide
- 239 tons of nitrogen oxide
- 43 tons of volatile organic compounds[45]

The plant produces about sixteen thousand tons of wood ash annually at 70 percent load factor. The utility expected to sell or give away the ash. However, a dump site was created at the plant site for disposal of any ash for which there was no demand. All ash produced is being used as envisioned as a soil amendment.

Despite the many positive features of the facility itself and the regulations under which it operated and harvested wood, the WFPP experienced a number of problems. Because of the scale and innovative nature of the project, these difficulties received regional and national attention including coverage in the *Wall Street Journal*. Many of the problems have been ascribed to the urban setting of the power plant. A rural site was considered but Burlington city officials demanded an urban siting in order for the city to receive payments in lieu of taxes that would come from the power plant. Constraints on the sale of power due to the dramatic fall in fossil fuel prices exacerbated problems at the facility.

The major problem was a large inventory of chips, both at the plant and the Swanton rail reloading facility.[46] The piles of chips began to heat up and smolder due to spontaneous combustion during the summer of 1985. Some rotting also occurred, creating odor and effluent problems. Actions taken to deal with this and to reduce the supply of chips resulted in the idling of nearly a dozen wood chip harvesters. Unplanned power was produced when the plant began operating around the clock to consume its wood chip inventory. This also helped to keep independent wood chip suppliers in business. Besides the dirt and odor coming from the piles of wood chips, noise from the plant has also been a persistent source of complaints from people residing adjacent to the plant.

The problems faced by the power plant are both economic and environmental. The heavily financed wood chip harvesters could go out of business without continued demand from the WFPP. The utility has been forced to produce power it doesn't need. At the same time the falling cost of fossil fuels has put all wood-fueled operations in a tenuous position. The difficulty may be short term but it is, nevertheless, a chronic concern for utility officials charged with producing power at the lowest possible cost. Because a viable wood supply is absolutely critical, dislocations in the wood market and their effect on landowners' willingness to harvest are a special problem for the utility. BED does not own any forestland and must obtain all of its fuel wood on the open market. Burlington twice passed a referendum to build the plant but may now have second thoughts about that support. The noise, odor, smoke, and effluents from the operation have created a legacy of public disenchantment.

Developers in Vermont have already found the going to be tougher. The proposal by Decker Energy International to site an 18-MW wood-fired power plant at Randolph, Vermont met with entrenched citizen opposition. Decker Opposition Association soon formed to protest the proposal. Much of the criticism proclaimed by the group was based on the problems at the Burlington plant. Decker soon abandoned this site.[47]

Perhaps the most significant effect of the Burlington experience will be felt well beyond the state of Vermont. Some observers credit news of Burlington's problems as causing opposition elsewhere in the United States.[48] Fortunately or not, the 50-MW facility stands as a demonstration of the use of wood to produce electric power. This use of wood energy is at an early and somewhat delicate stage of development. Other wood-electric developers around the nation will undoubtedly have to offer a defense pursuant to the bad news from Burlington.

WFPP DEVELOPMENT ELSEWHERE IN THE NORTHEAST

The greatest interest and the most intense development of wood for electric power has occurred in the northeast United States. The 50-MW Burlington WFPP is indicative of this fact. Five WFPPs with a total generating capacity of 60 MW came on line in New Hampshire alone during 1988. By the end of that year Maine had 230 MW of wood or wood/coal cofired capacity and projected an additional 250 MW by 1992. Seven other WFPPs fired with recycled wood wastes were in preconstruction phases in Connecticut, New York, Pennsylvania, and Massachusetts in 1988.[49]

New Hampshire is far and away the leader of this trend in the Northeast. In 1987 six biomass cogeneration facilities were operating there, producing 50 MW of power based on consumption of eight hundred sixty thousand green tons of wood residue. Partly due to favorable rate guarantees for small power producers, twenty-three applications for construction of WFPPs were received by the New Hampshire Public Service Commission that year. Eighteen of these plants were granted permits by the end of 1987. Should all of these plants be built, together they would produce an additional 160 MW of power for the state and annually consume 2.5 million green tons of wood.[50] By late 1988 New Hampshire had a total generating capacity of 140 MW in eleven WFPPs. These plants produce about 10 percent of the state's electricity needs. However, this achievement has not come without problems. About 70 percent of the wood utilized in these plants comes from landclearing operations.[51] Much of this has occurred downstate. The source of fuel after this temporary boom in production and the method by which it will be harvested are a source of concern. The potential conflict experienced in other WFPP controversies is also recognized for New Hampshire. For example:

> Ironically, the same people who are, directly or indirectly, footing the bill for residential and industrial landclearing in the southern part of the state, are those who most threaten the future of both biomass and conventional timber harvesting in the entire state of New Hampshire. Many of these people are new residents of the state—urban dwellers who have made good and want a more relaxed, rural, and aesthetically pleasing lifestyle.[52]

A backlash in the form of additional regulation is feared if and when this conflict develops. One observer also notes that the New Hampshire plants, when viewed on a map, are geographically clustered and therefore likely to concentrate whole tree chip production. The resulting competition for wood chips could become a severe detriment to the forest. Vermont has effective regulations for whole-tree chip harvesting and procurement, but New Hampshire has no such controls. This fact combined with the boom in WFPP development points up the potential for harm to New Hampshire's forests.[53]

Maine is also hotbed of WFPP development. More than twenty such projects were planned for the state as of 1985. If all are built, these plants would produce 479 MW and would annually consume about 6.3 million tons of wood.[54] Three of these proposed plants are 22-MW facilities to be developed by Ultrapower Services, Inc. of Irvine, California. Eight plants with a combined potential output of 207 MW were expected to be under construction in 1986.[55] The news note reporting this activity listed some of the issues for both promoters and opponents:

> Concerns are being raised, however, about the state's ability to produce a sufficient amount of biomass and mill residue. Opponents cite the unknown effects of nutrient drain and the potential for indiscriminate cutting prompted by year-around, steady demand for fuel.

Proponents ... maintain, however, that "waste" wood using plants can promote better forest management, cut down on waste in the woods, and provide new markets for sawmill residue.[56]

Two 25-MW WFPPs, located in Jonesboro and Enfield, started producing power late in 1986.[57]

The potential for natural resource conflict became a reality in Maine during 1987. A WFPP planned for Eustis, Maine, projected to cost $73 million and produce 40 MW of power, became the object of citizen action. Two lawsuits were filed by Friends of the Maine Woods and Friends of Bigelow to block the project. The Maine Department of Environmental Protection, the Maine Board of Environmental Protection, and Stratton Energy Associates, the project's developer, were all named in these lawsuits. The first suit alleged that the Board of Environmental Protection had violated the state's freedom-of-access law when it met privately rather than in a scheduled open public hearing. The second suit accused Stratton Energy Associates and the Department of Environmental Protection of failing to adequately consider the environmental impact of the project on the nearby Bigelow Mountain Wilderness Preserve. The Maine Site Location Development Law calls for such a process. Notwithstanding these lawsuits, considerable local support existed for the project, a response to the promise of additional jobs for the community.[58]

Planning continues in Maine for a set of smaller WFPPs. Six plants with a generating capacity of 15.3 MW each were on the drawing board for Maine's Aroostook County in 1989, despite the prospect that the projects might be killed by the purchase of hydropower from Quebec. The developer is Alternate Energy, Inc. The plants would be built in the Oakfield-Houlton area and at Van Buren, Ashland, Fort Kent, Eagle Lake, and Patten. Fuel for the $150 million project would come from whole-tree-chipping operations and sawmill residues.[59]

Vermont, of course, possesses the nation's largest WFPP at Burlington and produced as much power from this one plant as all New Hampshire did from its six WFPPs in 1987. However, perhaps because of unfavorable experiences with the Burlington plant, Vermont has been much slower to develop additional WFPPs. The state has enacted strict environmental safeguards. The only other plant to be developed has been awash in controversy. The Decker Energy International WFPP was turned away from the town of Randolph by intense local opposition in 1986. It is now planned for Ryegate, where the 18-MW facility was unanimously approved by the Town Planning Commission.[60] In view of the issues raised in this book, it would be worthwhile to study these two contrasting communities in connection with their decision on the plant. Although few privately owned WFPPs for the production of electricity are now planned for Vermont, a number of existing manufacturers have plans to establish such facilities on site for cogeneration.[61]

Public concerns have embraced proposals to site WFPPs in the state of New York as well. In 1985 Ultrapower Services, Inc., proposed to build two WFPPs in the upstate New York communities of Boonville and Gouverneur. The reception there was controversial, with some fearing that the forests would be depleted in order to generate power while others were enthusiastic about the prospective effect on the local economy and on forest management practices.[62] Delay in these and two other plants proposed by Ultrapower resulted from a decision by Niagara Mohawk Power Corp., and ratified by the New York Public Service Commission, to offer a variable-rate rather than a fixed-rate price schedule for electricity from small producers.[63] These plants were shelved late in 1986 as a result of the impasse and Ultrapower's disagreement with state policy on alternative energy development.[64]

Three Adirondack communities in northern New York—Malone, Edwards, and Tupper Lake—each planned to build a 20-MW WFPP as of 1988. The developer, New York City's Long Lake Energy Corp., hoped to have these facilities in operation by January 1990 in order to receive several tax breaks. The cost of each plant was put at $25 million. But unlike New Hampshire and Vermont, New York has a history of dashed high hopes, controversy, and false starts.[65] It remains to be seen if large-scale wood energy development can succeed in New York.

This account of WFPP development and some concomitant controversies in the Northeast has been pieced together from accounts appearing in trade journals. Such controversies are usually local and seldom reach the pages of larger newspapers or magazines. To know about them in any detail almost requires residence in the affected communities, a serious barrier for anyone seeking to survey wood energy controversies nationwide.

This experience with WFPP development in the Northeast brings up the question of why New Hampshire, by contrast with most other places in the United States, has so readily accepted WFPPs. Does it have something to do with migration patterns or local politics peculiar to rural New Hampshire? The lack of development in New York, like Minnesota and Wisconsin, may in part be due to the level of political sophistication among their rural populace. Both Maine and Vermont exhibit the potential for opposition to WFPPs that threaten environmental values.

If anything, the diversity of these responses points to the need for additional comparative study. Conflict over large-scale development of wood for energy in nonmetropolitan areas would seem to be a harbinger for similar development in the future. There is little reason to doubt that pressures for development will increase along with the growth of population and economic activity and the depletion of nonrenewable resources.

6

Policy Implications for Large-Scale Wood Energy Development

Wood Energy Policy in a Changing Context

The situation today is more complex than when the policy gap was identified between the need to produce energy and the need for protection of the environment.[1] Although the policy gap remains, citizen activity has continued to refine the terms of the debate. Scale and centralization have become the nexus of debate about the direction energy policy should encourage. Energy development in general and the siting of large-scale wood energy facilities in particular should be of increasing interest to policymakers as fossil energy resources are depleted.

The results of this study are germane to the policy debate about scale and centralization in energy development. Public concern about environmental quality and especially the hazards of power plants continues unabated. This concern extends even to large-scale renewable energy resource development in which the environmental risks are presumed to be of a lower order of magnitude.

In the controversies described in this book the public did not object to the small-scale uses to which the resource had been put, particularly home heating. Indeed, the threat to these uses was an important motivation for WFPP opponents. By implication, energy policy that recognizes the growing interest in small-scale energy technologies would seem to be more responsive to public concerns about wood energy development.

Opposition to WFPPs appears to be related more to the scale of the facility than to the characteristics of the resource. Scale is relative in the sociopolitical accounting of an energy source—WFPPs being a small fraction the size of conventional coal and nuclear power plants. The common denominator is the extent to which scale imposes a threat to environmental values. If wood-electric power is to be developed at all, then the scale of power plants will probably have to be downsized to be environmentally and thereby politically acceptable. Dispersed power plants 5 MW in size may be an appropriate limit.[2] It remains to be seen whether developers will accept the feasibility of building WFPPs at this scale.

Feasibility, of course, depends on prices for conventional fuels, which are certain to rise. Yet the idea of developing wood for power may be a passing fancy. Burning wood for electric power or even for cogeneration is a low-grade use of the resource. Residential wood burning, for example, is much more energetically efficient. The rise of biotechnology industries and the growth of the forest products industry may eventually displace the use of wood for electric power.

Part and parcel of public concern about the scale of energy technology is the distribution of costs. In this context the options in deployment of energy technologies appear to diverge on the basis of environmental risks. Small-scale decentralized technologies, such an burning wood for home heating or installing a solar collector on the roof, are qualitatively different from the risks of large-scale energy facilities. In choosing small-scale technologies the individual consumer generally assumes his/her own risks. The social choice for large-scale energy development, on the other hand, requires opponents to acquiesce to risks they do not wish to assume.[3]

An Assessment of Michigan's Wood Energy Policy

The actors in controversies seldom possess the kind of information about strategies of expertise stated so explicitly here. Moreover, there is an element of hypocrisy in the use of experts. On the one hand, they are held up as unbiased consultants whose only interest is to impart or apply knowledge. On the other, experts serve their essential purpose by depoliticizing technical decisions while decision makers ritualistically ignore the experts' role. Given the foregoing understanding of wood energy development controversies, it is altogether appropriate to assess Michigan's wood energy policy for evidence of the strategies of expertise.

This account looks critically at wood energy development policy in Michigan. Such policy presumably might have been improved by a better understanding of the technical controversies discussed in this book. The assessment proceeds from an examination of forest policy in general to consideration of the recent statement of policy promoting continued large-scale wood energy development. The

Typical second growth forest in the cutover region of the Great Lakes. Pictured are mixed hardwoods in northwest Michigan suitable for producing wood chips to fuel a power plant.

most important aspect to this review is the revelation that state agencies continue to use technical expertise to promote a controversial policy. The language used by policymakers suggests that deference will be paid to environmental and natural resource issues while the apparent encouragement of large-scale of development suggests quite the opposite. Because the issue of scale is not explicitly treated, the policy fails to address the environmental problems of scale that are the essence of public opposition.

This evaluation should not be construed as an argument for or against wood energy development. Rather, it serves as an illustration of the strategies of expertise. Michigan is cast as a proponent of technological change by virtue of its stated policy. Its position has already been found wanting at places like Hersey and Indian River. Nonmetropolitan communities in Michigan and elsewhere may eventually welcome wood-fired power plants, or they may unsuccessfully oppose them as happened at Westwood, California. Nevertheless, the strategies of expertise will again be played to their inexorable conclusion whenever implementation of such policies threatens critical social values.

The Initiative for Economic Development of
Michigan's Forests

Like Governor William Milliken before him, Michigan Governor James J. Blanchard has promoted policies aimed at economic development of the state's abundant forest resources. As mentioned earlier, Governor Milliken had convened a conference on industrial use of wood energy at the University of Michigan in November 1977 to respond to the energy crisis. Governor Blanchard followed suit, sponsoring a conference at Michigan State University in March 1983 on economic development of Michigan's forests. Each conference was an attempt by state government to deal with a crisis. The energy crises of the 1970s gave way to the state fiscal and economic crisis of the early 1980s. The economic recession was especially hard on the automobile industry. High unemployment and out-migration were the symptoms of a malaise Michigan sought to cure. Governor Blanchard, a Democrat, was elected with help from organized labor. In view of this fact and the state's soaring unemployment, it was entirely logical that he should subtitle the conference "Creating 50,000 New Jobs in Michigan Forest Products Industries."[4]

The latter conference encompassed the gamut of possibilities in economic development of forests. Wood energy was only a small segment of the conference agenda. It was not specifically mentioned in the governor's address or the recommendations of the conference. Two paragraphs on wood energy appear in an appendix to the conference proceedings on page 70 under the title "wood can supply more energy for home heat and generation of electricity."[5] A few research needs in wood fuel use and wood plantation technology are listed there.

The conference was just the beginning of a major initiative by the state of Michigan to develop its forests. Subsequent study and recommendations by executive agencies formalized policy.

An interim report in May of 1985 devoted two of its twenty-seven pages to wood energy development. The report briefly summarized fuel wood consumption in Michigan, covering both some positive and negative aspects. It concluded that "potential negative impacts can be avoided or minimized if steps are taken to insure that development of wood energy use occurs in an orderly manner with appropriate controls."[6] The major policy statement, soon to be released as the *Michigan Wood Energy Plan*, is mentioned, along with a description of the Michigan Biomass Energy Program.

Planning for Michigan forest development has taken place largely under the auspices of the Michigan Department of Natural Resources in the Stephen T. Mason Building in Lansing, Michigan.

The Michigan Wood Energy Plan

A sweeping policy statement on development of Michigan's forest resources finally appeared in 1986. An addendum to *Michigan Forest Resources—A Statewide Forest Resources Plan* dated March 1986—the *Michigan Wood Energy Development Plan*—was intended to define state policy on wood energy development. It had been created with the collaboration of all relevant state agencies and represented the current status of the understanding of and goals for wood energy development in Michigan government. The following discussion summarizes and reviews the plan. The review is informed by the foregoing study of Michigan's WFPP controversies.

One of the four major goals outlined in the *Statewide Forest Resources Plan* approved by the Michigan Natural Resources Commission is the improvement of Michigan's energy situation through energy responsive forest management. An ad hoc task force of both the Department of Commerce and the Department of Natural Resources developed the *Michigan Wood Energy Plan* to promote "an orderly development of a portion of Michigan's wood resources for energy production that is sensitive to multiple-use and environmental concerns."[7] The plan proceeds from the goal of influencing the market for and supplies of wood fuel while allowing the market to determine the best and highest value use for wood. Draft copies were circulated among individuals, organizations, and agencies in Michigan. Meetings intended to provide public input were conducted statewide. Comments in meetings and letters reflected the concerns voiced in the foregoing controversies. "All suggested improvements were incorporated into the plan where feasible," according to the report.[8] The plan was approved by the Michigan Natural Resources Commission on February 6, 1986.

Michigan's energy situation is defined as part of a background summary in this document. Some attention is given the role that wood energy development could play in meeting the state's energy needs. The plan covers current consumption, wood's cost advantage relative to other fuels, use of wood for energy in the context of a forest product industry development planning process, the relationship between wood use and jobs, and the salutary effect that markets for "waste wood" might have on improving forest productivity, wildlife habitat, and aesthetic management of wood harvesting. The background summary concludes that using wood for energy presents opportunities to:

- Decrease energy costs for Michigan homes, businesses, and institutions
- Reduce the drain of Michigan dollars for imported energy
- Provide jobs in harvesting, transportation, construction, and facility operation
- Increase incentives to landowners to practice forest management for timber, recreation, wildlife, and aesthetics by providing markets for low-value wood.[9]

The general purpose of the plan is stated as follows:

... to provide direction to and improve coordination of state agencies involved in wood energy and *to provide public input into this process*. It is intended that this plan will facilitate an orderly development of a portion of Michigan's wood resources for energy production that is sensitive to multiple-use and environmental concerns. [Emphasis mine.][10]

Within this framework specific goals and activities are listed and described including wood supply and utilization, economic development support, environmental considerations, and wood energy facilities. Table 10 reproduces goals and activities related to

The corridor of executive office buildings directly west of the capitol in Lansing, Michigan, a center for administrative decision making in the state.

the environment. The report provides a series of tables summarizing these goals and activities. It is a technical as well as a policy document, written with the assistance of technical experts in the employ of the state of Michigan. An assessment of the policy as articulated by these goals and activities is a worthwhile exercise for putting the foregoing study to work. In this spirit, the *Michigan Wood Energy Plan* and the policy it states are reviewed here in terms of citizen involvement.

A forthright statement of policy on the scale of development to be encouraged is not part of the document. It does not explicitly address the relationship between scale and environmental impact. Concerns demonstrated or expressed by citizens in areas where WFPPs have been proposed are thereby ignored. Instead, a mixed bag of goals and activities is offered to promote development of Michigan's forests for energy without proper specification of the direction of development and probable environmental impacts (see table 10). Development that assures minimal environmental impact is suggested, but no research evidence on the nature of known impacts is provided. Impacts are, in turn, a function of technological scale, a relationship that has been clearly recognized by opponents of the technology.

TABLE 10

Michigan Wood Energy Plan **Environmental Considerations**

Goal:	To ensure that the development of wood energy does not sacrifice environmental quality or the balance of ecological relationships.

Activities:

- Develop a program which will (1) identify current research on the effects of whole tree harvesting on erosion, soil nutrient depletion, damage to residual trees, forest regeneration, the effects of snag and down timber extraction on wildlife cover, avity nesting and food sources, and other ecological relationships impacted by whole tree harvesting, (2) identify research results applicable to Michigan forests, (3) determine areas current research does not address, (4) develop a strategy to coordinate and support research to meet the needs of Michigan forest managers, and (5) implement an action plan to prevent environmental degradation.

- Monitor air and water quality at wood energy facilities and develop specific operating standards in accordance with the Michigan Air Pollution Act, P.A. 348, 1965, Michigan Water Resources Commission Act, P.A. 245, 1929, the Federal Clean Air Act Amendment, PL 95-95, 1977 and the Federal Clean Water Act PL 92-500, 1972 as amended. Review existing air quality statutes and regulations pertaining to burning wood for energy to assure the protection of public health and well-being.

- Develop a research based action plan which addresses the risks of residential woodburning in the following three areas: (1) indoor air quality, (2) ambient air quality, and (3) home safety.

Source: State of Michigan, *Michigan Wood Energy Development Plan: An Addendum to Michigan's Forest Resources, A Statewide Resources Plan*, Department of Commerce and Department of Natural Resources, Lansing, March 1986, p.5.

It is arguable that keeping policy general better accords with large-scale development than with small-scale development of wood for energy, but a case can also be made that this amounts to no policy at all. Making policy sufficiently general and giving assurances of future consideration for environmental hazards leaves the door wide open for controversial developments. Of course the intention is not that the developments be controversial, just that they take place in order to achieve policy goals. Citizens in northern Michigan will doubtless be interested in where the policy is leading particularly if their community is chosen as a site for a WFPP. They are likely to be interested in the effect on their local resources if called upon to supply fuel for the facility or the impact on their environment if required to live with pollution from the WFPP. In energy development policy it is increasingly necessary to respond to underlying value issues by explicitly tackling the issue of scale. Ignoring values as irrelevant to the technical dimension of energy policy making sets the stage for conflict when the environmental implications of scale become known to the local community.

Similarly, concern is expressed in the plan for coordinating the activities of state government, but none can be found for the interaction between state and local government.[11] This presumes that a policy can be defined and implemented entirely by means of state government with little or no interaction with local communities or their elected officials—at best a naive view of how technological issues are decided. Local government is a nexus of political activity. Large-scale projects can succeed or fail when local communities confront the issue of environmental impacts. Dissenting citizens have learned that this is the level of government where they can have the greatest effect.

Whatever the values at stake in technological controversies, conflict inevitably gets to the matter of local autonomy. The state is perceived as a distant, centralized, and unresponsive bureaucracy not necessarily concerned with the local community. Such proposals as might be encouraged by the Michigan wood energy policy represent a potential threat to local autonomy. It is not sufficient to recognize the decision making authority of local government. State government must consider issues from the point of view of local government in order to enhance the prospect that the policy will succeed. This is not necessarily to suggest further sharing of power with local government, merely that the success of a policy hinges on what is socially and politically feasible. If a policy is unlikely to be ratified at the local level, then it would seem to be ineffective public policy from the start. No where is this more evident than in the siting of large-scale facilities.

The deference paid citizen participation in the *Michigan Wood Energy Plan* is typical of state resource policy and management practices. This may be the most intractable part of the stated policy. Public input is not the same as citizen involvement. The evidence presented here demonstrates that activities to involve citizens in decision making are often token. In each controversy public hearings led to conflict rather than consensus. There is little evidence in the plan that citizen participation in the form outlined had any impact on policymaking, save for several weak proposals to insure that environmental impact is acceptable.

The essential problem is that a method of citizen participation has not been articulated. Soliciting public comments at hearings may be useful for plumbing the depths of public concern, but it by no means represents a public voice in the decisionmaking process. The tendency to rely on technical experts for decision making is almost irresistible in government bureaucracies; but once again, this pits technical experts against citizens in matters of values as well as facts. Furthermore, the dimensions of citizen concern generally do not become clear until a policy is implemented at the local level. It usually requires a concrete proposal that affects *your* community for the value issues to surface. From this perspective hearings on wood energy policy that do not consider the elements of technological scale and local autonomy are flawed from the start. The resulting policy risks complete disaster when the time arrives for implementation.

An example should serve to illustrate this point. Federal law enables the states to enter into compacts for purposes of disposing of low-level radioactive waste. In doing so a group of states can select one of its members to be the initial site for a disposal facility. On the face of it this seems like a happy solution to the problem. The full power of the federal and state governments has been brought to bear on the growing problem of radioactive waste disposal. Citizen participation has been thoughtfully integrated into the process, with public hearings scheduled at every stage of the site selection process. There is, however, quite another side to this solution. When citizens are asked to participate in the initial rounds, there is very little controversy—a situation comparable to the hearings held in connection with development of the *Michigan Wood Energy Plan*. But when the process gets down to picking the actual site of the facility, then controversy erupts with a vengeance. Now radioactive waste disposal is an unproven technology with serious hazards and uncertainties. There is ample room for disagreement among experts. If citizen participation is not genuine, this is the stage at which siting radioactive waste facilities will be defeated. In a similar fashion, additional attempts to develop large-scale wood energy facilities under the rubric of the Michigan wood energy policy is likely to fail without genuine citizen participation. If citizens are not heard and heeded in one forum, they will find others that are usually more adversarial. If state administrators lend a deaf ear, then the politics

176 WOOD ENERGY CONTROVERSIES

of protest and responsiveness of local government are always available.

The *Michigan Wood Energy Plan* has been briefly summarized and reviewed as an example of a three year effort at making policy gone awry. The authors of the plan fail to confront the problem of citizen acceptance of technological development. On top of the environmental and social concerns revealed in defeats at Hersey and Indian River and ignored in this plan, Michigan officials promoting WFPPs must contend with the widely reported problems of the 50-MW Burlington WFPP. At the same time conflict over increased logging and growing competition from wood processing industries make the specter of WFPPs in all corners of northern Michigan an unlikely possibility.[12] Consider as well that rising fossil fuel prices will probably lead to use of forest products as a replacement for more energy intensive materials, thus accelerating conflict with wood for energy.[13] This study would predict that, to the extent the goals and activities of the plan promote large-scale WFPPs, there will be continued conflict over the development of wood from Michigan's forests for electric power.

Private development is succeeding where government-sponsored efforts failed. The strategy at work here seems to be one of siting and building WFPPs without regard to administrative decisions about wood allocation from public lands. The corresponding lack of public decision making means that citizens have much less opportunity to take part in the decision, even if their values are threatened. The trend in site selection is apparently toward remote, rural communities lacking the political diversity to mount a successful opposition. In most cases these communities are already familiar with wood processing operations. Indeed, the major source of fuel for the power plants is sawdust and other wood wastes—making this development comparable to the widespread use of wood wastes for fuel in the forest products industry of the western United States. Similarly, the Dow Corning SECO plant was sited in a community familiar with and favorable to industrial development. Siting facilities in economically depressed areas, away from scenic and recreational amenities, is another factor in the successes reviewed here. Development was arguably consistent with local values connected with a desire for the local benefits of economic development.

The size, experience, and public perception of the developers appear to have had little impact on the reception of WFPPs. Both Dow Corning and Consumers Power have less than desirable reputations in matters of environmental quality. Consumers received a lot of bad press over its nuclear power plant, but this was incidental in the Hersey controversy where the conflict was directed at government. Yet Dow Corning succeeded in building a WFPP and Consumers failed. By the same token, Primary Power met defeat in its bid to site a smaller plant at Indian River, whereas Viking Energy Corporation is succeeding with the WFPPs it is building at Lincoln and McBain. The social variables described in this study, most especially political diversity and environmental values in the affected communities, better account for these contrasts.

Using the strategy just described, private developers persuaded the communities of Hillman, Lincoln, and McBain to accept WFPPs. It has been suggested that similar power plants will dot the northern Michigan landscape in the 1990s.[14] However, further development will inevitably bring up the issue of allocation of wood from public forests, exacerbate conflicts with other uses of wood, and affect the quality of the environment for recreation and tourism. This study suggests that the success of this development effort ultimately depends on citizen involvement in the decision making process. If government does not act to involve citizens, then the pattern of citizen participation observed at Hersey and Indian River is likely to be repeated despite Michigan's wood energy policy and because of its shortcomings.

Part III
A Social Study of Wood Energy Controversy

7

Values, Facts, and Ideology
in the Hersey Controversy

Conflict predicated upon diverging values is the fundamental basis for technical controversies. Although technical disputes frequently revolve around matters of fact, in the end they entail political choices among competing social values. Obvious as this may seem, social scientists have traditionally paid little attention to the interplay of facts and values in controversies. This neglect has changed for the better. A number of scholars have taken up the study of facts and values especially in connection with practical environmental concerns.[1]

How are matters of value different from matters of fact? Values are " those conceptions of desirable states of affairs that are utilized in selective conduct as criteria for preference or choice or as justifications for proposed or actual behavior."[2] Facts are conceptions that can be objectively verified. They refer to "what is," whereas values concern what is good or desirable. Values determine "what ought to be." The word "normative" is used to describe notions of what is good or desirable. Technical decisions have normative as well as factual content.

David Hume, the philosopher, suggested the logical distinction between "what is" and "what ought to be." The importance of this distinction for the study of technical controversies is that experts can help establish what is, but they have no special qualifications for ascertaining what ought to be. In a democratic society that, in principle, is the business of the people.

Values and facts are first order concepts—thus better illustrated than defined. In many siting disputes the prospect of additional jobs becomes an issue. The number of jobs created and their nature (for example, whether locals or outsiders will get them) are matters of fact. Conflict may occur over whether one or another forecast is accurate or the extent to which unskilled local workers can be hired—that is, over the facts—but in most cases values are the crux of the debate about jobs. For example, the good of additional jobs relative to other value concerns may be questioned. However, few would deny that additional jobs represent a positive value unless the jobs go primarily to outsiders or to transients who depart after a socially disruptive construction phase. In practice the value of the prospective jobs to the community is weighed against the social and environmental costs of the proposed facility. Risks and benefits may also enter the debate as factual issues, but they, too, are increasingly gauged in terms of social and political values.[3] Evaluations about what is desirable or what the community should do are not subject to the methods that establish the facts.

Experts may help bring forth the facts, but they are not particularly useful in ascribing values. Indeed, experts generally claim their only concern is the facts and nothing but the facts. This is a norm that the practitioners of science are expected to adopt when they join the profession. Logical positivism is strict on this point. Yet the claim of neutrality among scientists is itself ideological. According to Mulkay, this position is tantamount to a selective employment and interpretation of cultural resources available to scientists.[4] Abjuring values has served as a means to establish the boundaries of science, isolating it from politics and religion. This has permitted science to flourish in the domain where its methods have been most productive. However, the persistence of this norm in the social sciences, which have long sought to emulate the natural sciences, presents a number of problems that go beyond the scope of this study. Suffice it to say that the subject matter and results of social science research have value implications that are difficult to ignore.

In both the social sciences and the natural sciences, decisions about values begin with the choice of problem, the method used to study it, and the level of precision expected in the results. Even these decisions often prove controversial when disputes arise. Barbour finds six reasons why experts can disagree:

- Uncertainties in data
- The way the issues are defined
- Judgments of value
- Professional biases
- Institutional biases
- Conflicts of interest[5]

Notwithstanding the value choices of scientists and technicians, the social values of the parties in the conflict are the ultimate source of controversy. Great care must be taken to observe or impute these values when studying technical controversies.

Controversies graphically show that science is anything but objective in the decision making process. Project promoters very deliberately make use of expertise to legitimize their plans. They employ experts and rely on command of technical knowledge for the purpose of justifying their (the project promoters') autonomy. Important value-laden decisions are, in effect, defined as part of the job to be accomplished by experts. For example, Feldman and Milch found the results of technical studies to be consistent with preferences of sponsors for three basic reasons: (1) experts were selected on the basis that they shared a common perspective with their employers; (2) the sponsors defined the terms of inquiry at the outset, thus limiting the definition of the problem; and (3) long term economic interests dictated that experts provide results according to client preferences or fail to get contracts.[6]

In response, "local experts" are increasingly called upon to counter establishment experts. Technical knowledge functions as a tool exploited by divergent interests to reinforce their respective claims. Conflict between experts shifts attention to nontechnical and political assumptions—that is to say to values. Intensely held values lead to a reduced role for "knowledge as an impartial arbiter." Technical expertise is now widely recognized as the only way to challenge controversial decisions, whatever the motivating political or moral values.[7]

Values by themselves do not produce conflict. Knowing what is desired or good does not imply a corresponding course of action. Individuals and groups must also know how to obtain values. This is where ideology comes into the picture. Ideology provides a framework containing both a coherent set of values and a program for achieving them. It serves to define a plan of action. In order to form groups for purposes of promoting or opposing a project, an organizing ideology must first be present. By implication, the quickest way to catalog the values in a dispute is to find a statement of the ideologies present.

The ideologies of the hard path and the soft path in energy development discussed in chapter 3 find clear and convincing application in the Hersey controversy. The following analysis of facts, values, and technical expertise in the Hersey controversy looks at the interactions of fact and value in policy decisions in arenas where professional expertise plays the lead role. It shows how particular values mesh with the ideologies of contending parties in the controversy and dictate the course of the debate.

Sources of Data and Information

The media typically become a forum for debate in controversies.[8] Very often these accounts are finely detailed, bringing out the full range of facts and values in the controversy as well as the ideologies that influenced public debate. Using this kind of data therefore permits study of a controversy as a whole.

The *Osceola County Herald* (*OCH*)—a weekly newspaper published at Reed City, the county seat—quickly became the lightning rod for the Hersey WFPP controversy. The *OCH* contains a wealth of data for study of the controversy. A total of seventy-eight feature articles, columns, and letters to the editor about the WFPP appeared in the *OCH* during the controversy, beginning in September 1978 and concluding in September 1980.

Other newspapers published articles on the controversy, but their coverage was far less frequent and much less detailed. This observation is consistent with the finding that the quality of coverage goes down with increasing distance from the affected community.[9] Accounts of the Hersey controversy apppeared in the *Evart Review* (a weekly newspaper published in a small town just to the northeast of Hersey), the *Big Rapids Pioneer* (the newspaper for the nearest town of any size, located just to the south in adjacent Mecosta county), the *Grand Rapids Press* (a regional newspaper serving western Michigan), the *Lansing State Journal* (published in the capitol city of Michigan), and two newspapers from the largest metropolitan area in the state—the *Detroit News* and the *Detroit Free Press*.

Relevant articles from these newspapers were not analyzed but have been used to check on the validity of findings from the analysis of the *OCH* accounts. Similar use was made of articles in two Michigan magazines, namely *Michigan Natural Resources* (a bimonthly, glossy magazine published by the Michigan DNR) and *Michigan Out-of-Doors* (a monthly magazine of the Michigan United Conservations Clubs, also published in Lansing).

The following list shows the distance-decay function of media coverage in the controversy. Distances are from Hersey to the city where the newspaper or magazine is published:

- *Osceola County Herald*—78 accounts, 5 miles
- *Big Rapids Pioneer*—45 accounts, 14 miles
- *Grand Rapids Press*—13 accounts, 75 miles
- *Lansing State Journal*—7 accounts, 121 miles

- *Michigan Out-of-Doors*—6 accounts, 121 miles
- *Michigan Natural Resources*—2 accounts, 121 miles
- *Detroit News*—3 accounts, 203 miles

Another data and information source that captures the range of concerns about the proposed WFPP is the set of public hearings held prior to selection of the Hersey site. Transcripts of each were appended to the feasibility study for the WFPP.[10] This is a particularly interesting source because of the way the hearings were scheduled. An initial hearing specifically intended to ferret out environmental concerns was held at Big Rapids on 25 September 1978. Hearings quickly followed at or near each of the three prime sites. The last hearing was held at Reed City to discuss the Hersey site. Each of the prior hearings had been extensively reported in the *OCH*. As a consequence, citizens who attended the Reed City hearing were in a position to know about concerns expressed at earlier hearings.[11] The soon-to-be local experts for CRUF were the most vocal participants from the Hersey community.

Analysis of the Reed City hearing provided a baseline for comparison with the main content analysis of the *OCH*. The transcript of this hearing was also used as an information source to help insure the validity of the primary content analysis.

In addition to using the supplementary sources noted above, the veracity (that is, accuracy of reporting) of *OCH* accounts was checked against the Daverman and Associates feasibility study and CRUF's position paper and other documents.[12] These sources also yielded examples to complement issues found in the case history. Conclusions reached on the basis of the content analysis were enhanced by the knowledgeable use of these information sources.

Post-hoc interviews with selected proponents, opponents, and newspaper reporters were conducted largely for the purpose of fleshing out the case history. However, this information was also selectively used to carry out and validate the content analysis. The interviews employed open-ended questions administered by telephone. Questions were raised about the effect of the Hersey controversy on the local community and its politics, the basic issues and how they changed or evolved, the technial arguments raised for and against the WFPP and the interests of those who proposed them, the effect of these technical arguments in the dispute, changes in public opinion during the controversy, the community perception of the technical experts and their impact, the lessons contained in the controversy, and how the PBB contamination may have influenced the dispute.

Fact-Value Content Analysis

"Content analysis, in its broadest sense, refers to any interpretation of the contents of written materials. Social scientists use the term to mean objective and systematic analysis of the symbols embodied in communications."[13] It may involve either of two basic approaches. First is the development and enumeration of categories that isolate specific words, phrases, or sentences. A second approach—the one used in this study—is to categorize ideas or themes in a way that permits their quantification. Most essential in this regard are the criteria for identifying ideas or patterns of meaning. Content analysis by this method can be both qualitative and quantitative.

The content analysis of facts and values in the Hersey controversy is based on two units—individual arguments and time. Each of the 721 arguments raised in the accounts analyzed was coded and scored according to a fact-value typology, as follows:

- *Strictly factual.* Technical argument with no apparent value argument

- *Factual normative implied.* Technical argument with an ascribable value argument

- *Factual normative apparent.* Technical argument conjoined with a value argument.

- *Strictly normative.* Value argument without a corresponding technical argument

The unit of measure was the "string of the argument," which in practice ranged from a few words to several paragraphs. A code was assigned to indicate whether an argument was advanced by a proponent or an opponent or by an expert or a nonexpert. The time unit selected was two-month intervals, a total of twelve for the controversy. In most of the tables that follow time intervals were aggregated into two major phases. Phase I covers intervals 1–5 and Phase II intervals 6–12. This bifurcation takes advantage of a natural break in the controversy at interval 5. Some relationships are more obvious when the data are summarized in this fashion.

For purposes of this research study, argument types are operationalized as a scale, a fact-value continuum if you will. Arguments were evaluated using the scale, and scores were derived from their sum.

Scores should be interpreted as follows. A strictly factual argument would rate a 1 on the scale. Its polar opposite, a strictly normative argument, would rate a 4. The median score on the scale is 2.5. This figure would not be "neutral" in terms of facts and values. The median score is assumed to be value-laden because even highly technical arguments generally reflect values in such terms as efficiency, effectiveness, economy, welfare, etc. Hence, the scale is designed to accommodate the apparent distribution of arguments. In any event, the scale is relative and thus capable of producing a description and evaluation of the arguments in the controversy.

A single issue can be used as an example of how the scale was implemented. The issue is the efficiency/equity of the WFPP. Examples of each argument type demonstrate the range of arguments to which the scale might apply. "The WFPP will be 25 percent efficient in converting biomass [chemical energy] into electrical energy" is a Type 1 argument. It is a simple statement of fact without comparison, evaluation, or value assertion. "The WFPP is inefficient compared to residential woodburning—25 percent versus 70 percent for an airtight wood stove" is a Type 2 argument. Although a statement of fact regarding efficiency, it implies that the higher efficiency of residential wood burning makes this technology preferable to the WFPP. The value concern is implicit. A Type 3 argument for this issue might be: "The WFPP is the least we can get for this wood; firewood and local logging are a better use of the resource." Factual in asserting alternatives, this argument explicitly comes down on the side of local, small-scale use. "It's wrong to take firewood from poor people so the utilities can make a profit" is a Type 4 argument. It clearly states a value preference.

The Reed City hearing of 24 October 1978 and the seventy-eight accounts in the *OCH* are the complete data set for the content analysis. Findings of the analysis are presented next.

Facts, Values, and Expertise in the Hersey Controversy

The 24 October 1978 hearing at Reed City on the Hersey site was a discrete event in the controversy. As such it is but a cross-section of a dynamic social process. It could hardly serve as the basis for judging the controversy as a whole even if it was taken as a measure of public support or opposition to the WFPP. On the other hand, the number and diversity of arguments raised at the hearing provided an indication of the form the debate might take. It was, therefore, deemed appropriate to analyze the hearing to establish the issues evident at the outset and to provide a base for comparison with the overall debate.

Analysis of the Reed City hearing is summarized in table 11. The distribution of fact-value scores fell nearer the low end of the fact-value scale. Values became more important during the controversy. Opponent nonexpert scores were also much lower in the hearing than in the controversy proper. These observations suggest that the hearing is a special case not capable of being generalized to the controversy as a whole. It was, however, a baseline indication of other tendencies in the *OCH* analysis. For example, the relationship between proponent and opponent experts seemed to hold true for both the hearing and the overall controversy. In each case proponent experts had a very low score (skewed toward "facts"), and opponent experts had a substantially higher score (skewed toward "values").

TABLE 11

Fact-Value Characteristics of Arguments in the Hersey WFPP Site Hearing of 24 October 1978

	Arguments by Experts			Arguments by Nonexperts			Totals		
	n	sum of ratings	score	n	sum of ratings	score	n	sum of ratings	score
Proponents	44	51	1.16	72	158	2.19	116	209	1.80
Opponents	40	95	2.38	24	46	1.92	64	141	2.20
Uncommitted	0	—	—	26	41	1.58	26	41	1.58
Totals	84	146	1.74	122	245	2.01	206	391	1.90

A caution is in order here that applies to the entire analysis. Fact-value scores should not be interpreted as representing value judgments about the worth of the particular approach taken or the specific issues debated. Indeed, just because something was debated in the arena of "facts" did not mean it was factual. Many inaccuracies and deceptions were bandied about with the label "fact" during the controversy. In short, "fact" does not imply good and "value" bad, and vice versa.

Uncommitted participants were very much in evidence at the Hersey site hearing (n = 26 arguments). The uncommitted played such a small part in the *OCH* data (n = 12 arguments) that, for practical and methodological reasons, the decision was made to delete them from the primary analysis. Of these twelve arguments, five were reported in the initial *OCH* account and were made by an explicitly neutral representative of the Michigan United Conservation Clubs. None of the issues raised by uncommitted parties were unanticipated or unrecognized by the proponents and opponents.

A number of interesting patterns are revealed by analysis of the debate in the *OCH*. General results are presented in table 12–14.

The temporal distribution of the debate presented in table 12 reflects changes in the amount of controversy. Three peaks in the debate are clearly revealed by these data. The first and greatest peak was in the January–February 1979 interval when conflict enveloped the announcement that Hersey had been selected over Harlan and Whitehall as the site for the WFPP. The flurry of activity on both sides at this time established the foundation for the controversy. The number of arguments raised in this interval was more than double that of any other in the dispute. The second peak occurred at the end of 1979. It corresponds to the period of intense citizen concern about the possibility that solid or toxic waste would be burned in the plant. The third peak appeared in the May–June 1980 interval when the utilities withdrew from the venture after failing to get a speedy response to their request for a permit from Hersey township to experimentally burn RDF.

TABLE 12

Temporal Distribution of Hersey WFPP Accounts and Arguments in the *Osceola County Herald*, September 1978 – September 1980*

	1978		1979						1980				Totals
	1	*2*	*3*	*4*	*5*	*6*	*7*	*8*	*9*	*10*	*11*	*12*	
Accounts	3	3	13	5	0	1	8	12	7	8	14	3	78
Arguments	52	29	201	77	0	6	57	99	33	58	78	31	721
Percent of All Arguments	7	4	28	10	0	1	8	14	4	8	11	4	100

*In this and subsequent tables and figures the 12 intervals are consecutive bimonthly periods beginning with September–October 1978 and concluding with July–August 1980. Note the two major peaks, one at interval 3 and the other at interval 8. Advantage of the natural break at interval 5 was taken to divide the controversy into Phases I and II for purposes of summarizing the data.

Further analysis shows that proponents and opponents are reasonably close in number of arguments made (refer to table 13). However, as the trend in table 14 shows, proponents dominated the debate early, with the opponents moving ahead during the rest of the controversy. Also evident in table 13 is the relatively greater participation in the debate by proponent experts. Opponent contributions were almost evenly divided between experts and nonexperts. It is significant that experts for the proponents seemed to hold sway at the beginning of the controversy, while experts for the opponents began to dominate as the dispute evolved. For both proponents and opponents, nonexperts gradually increased in importance during the controversy. Nonexperts carried more of the debate for both sides in the later stages of the controversy.

TABLE 13

Quantity and Source of Arguments

	By Experts		By Nonexperts		Total
Proponents	198	57%	148	43%	100% = 346
Opponents	183	49%	192	51%	100% = 375
Totals	381	53%	340	47%	100% = 721

TABLE 14

Temporal Distribution of Arguments

	Phase I		Phase II		Total
Experts	224	58%	159	42%	100% = 383
Proponents	164	82%	36	18%	100% = 200
Opponents	60	33%	123	67%	100% = 183
Nonexperts	135	40%	203	60%	100% = 338
Proponents	63	43%	83	57%	100% = 146
Opponents	72	38%	120	63%	100% = 192

Experts, as expected, were more "fact" oriented than are nonexperts (refer to table 15). The analysis suggests that, in general, proponents are more oriented to "the facts" and opponents to "values." Nonexpert opponents scored reasonably close to their proponent counterparts. Proponents' arguments more often fell toward the "fact" side of the ledger and opponents' arguments toward the "values" side.

TABLE 15

Characteristics of Argument Rated on the Fact-Value Scale

	Arguments by Experts			Arguments by Nonexperts			Totals		
	n	sum of ratings	score	n	sum of ratings	score	n	sum of ratings	score
Proponents	198	368	1.86	148	366	2.47	346	734	2.12
Opponents	183	471	2.57	192	508	2.65	375	979	2.61
Totals	381	839	2.20	340	847	2.57	721	1713	2.38

Over time, there was a marked tendency for expert contributions to the debate to shift toward the values end of the scale (tables 16 and 17). Experts of both stripes experienced this shift. In fact, the debate as a whole followed this trend. The shift occurred for opponent experts even though they started from a much higher base than proponent experts. Proponent experts seemed to wind up where the opponent experts began relative to the scale.

TABLE 16

Fact-Value Scores for Experts and Nonexperts

	Phase I	Phase II	Total
Experts	1.83	2.70	2.19
Proponents	1.74	2.27	1.84
Opponents	2.05	2.83	2.57
Nonexperts	2.51	2.63	2.59
Proponents	2.51	2.51	2.51
Opponents	2.51	2.73	2.65

Experts were, proportionately, a much more significant part of the controversy at its beginning. Table 16 suggests that they tended to stick closer to matters of fact at this stage. Note once again the tendency for values to become part of the repertoire of experts.

TABLE 17

Quantity and Quality of Arguments by Experts

	Phase I	Phase II	Total/Mean
Experts Arguments	224	159	383
Percent of Arguments in Time Period	62	31	53
Sum of Fact-Value Ratings for Experts	409	430	839
Fact-Value Scores for Experts	1.83	2.70	2.19

The quantity of debate by the participants can also be viewed in terms of participation by experts and nonexperts. During the debate there was a disproportionate rise in the volume of arguments relative to the contribution by experts. This relationship is better illustrated by figure 4. The growth of participation by nonexperts at the expense of experts is particularly evident in the second phase of the dispute. This result can be interpreted to mean that disagreements among experts will stimulate further debate, particularly by nonexperts.

The quality of the debate in terms of the fact-value continuum is illustrated in figure 5. Experts' fact-value scores, although generally lower, do not seem to be out of step with all scores. Experts may be following the trend toward normative issues rather than leading it. There was an increase in expert debate about values corresponding with the general tendency to debate value issues.

Finally, comparison of fact-value scores for proponent and opponent experts (figure 6) shows them to be fairly close, with opponent experts apparently being much more willing to incorporate questions of value into their arsenal. This implies that vigorous attention to values is an advantage in technical controversies. The success of the opponents in technical controversies may stem from their willingness to embrace value issues. This pattern of behavior is not inconsistent with Altimore's suggestion that proponents try to narrow the scope of controversy and to dispense with "intangibles"

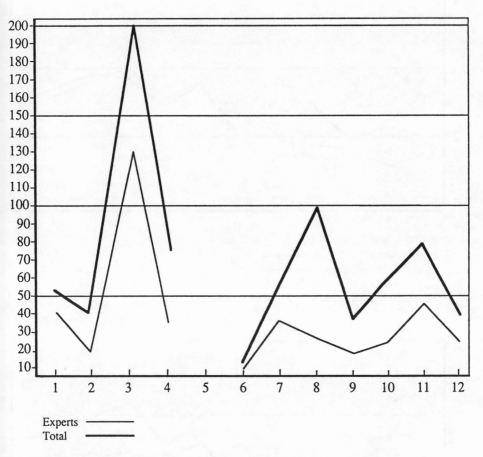

Experts ————————
Total   ————

Fig. 4. Quantity of expert arguments relative to total

whereas opponents attempt to broaden discussion.[14] The findings summarized in figure 6 imply that proponents do not succeed in this and are forced to deal with the value issues in technical controversies.

The Essential Connection between Values, Ideology, and Social Action

Experts very often do not recognize the role that they actually play in technical controversies. They consider themselves to be above ideology. They assume that their appeal to the norms of science is nonideological. But just as facts are difficult to distinguish from

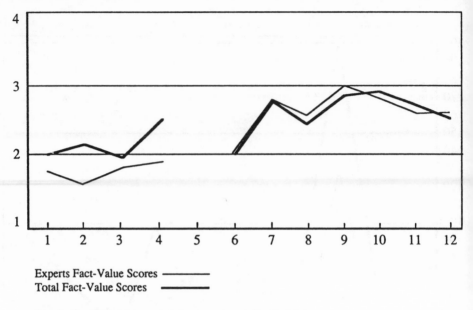

Experts Fact-Value Scores ——————
Total Fact-Value Scores ——————

Fig. 5. Quality of arguments by experts relative to total

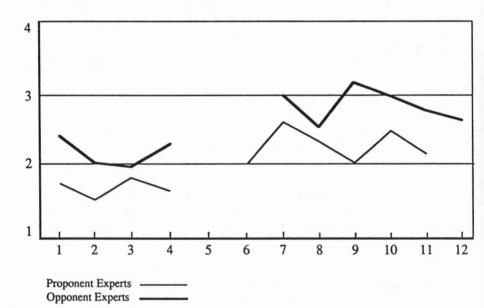

Proponent Experts ——————
Opponent Experts ——————

Fig. 6. Fact-value scores for arguments by experts

values in technical controversies, so, too, are the experts difficult to separate from the partisans. The experts involved serve a function whether or not they are aware of it. Their impact is determined by the values that give rise to the controversy. These values, in turn, are defined and mediated by ideologies.

Generally speaking, the ideology of developmentalists is opposed by the ideology of environmentalists in wood energy development controversies. The former values economic growth, centralization, and large-scale enterprise, while the latter seeks the values of environmental quality, local autonomy, and decentralized small-scale enterprise. Arguments couched in both cosmopolitan and local terms are brought to the debate by each side. Because the outcome is often determined by local government, the local perspective is favored and ideology is translated into locally-oriented arguments. The trade-off between economic growth and environmental quality is a difficult decision for local communities. However, in wood energy and other siting controversies, communities more often come down on the side of environmental quality. The technical debate that engenders this result is dominated by the experts and expertise of both proponents and opponents. That the experts disagree about the facts is less important sociologically than that they serve to defend a particular set of values as defined by their ideology. Understanding the role of experts and the importance of ideology is central to interpreting technical controversies, a topic to be further explored in the next chapter.

The Hersey case supports this view of the interdependence between values and ideology in technical controversy. Environmentally conscious individuals in the Hersey community, many of whom burned wood to heat their homes, viewed the proposal to burn wood for electric power production quite differently from the promoters of the project. These individuals saw an ideological difference between small-scale and large-scale energy development. Scale is dictated by the resource. For example, a WFPP cannot begin to approach the size of a typical nuclear power plant. Activists at Hersey thus took an ideological view of the threat to their environmental values constituted by the scale of the proposed WFPP.[15]

The issues of scale and centralization in energy development and especially in siting disputes were well demonstrated in the Hersey case. The choice came down to small-scale development of the resource for residential wood burning (in addition to continued small-scale logging) versus whole-tree harvesting on a large-scale to supply fuel for the WFPP. Notwithstanding the fact that wood is a renewable energy resource, this is fundamentally a choice between soft path and hard path development. Both proponents and opponents explicitly recognized the ideological debate and used familiar arguments to promote their side in the dispute.[16]

Examples from the debate show the importance of ideology in the controversy. An opponent of the WFPP proposal at the Hersey site hearing clearly articulated the case for a different technology (that is, the soft path) for utilizing the resource:

> You're here to find out what our options are too. Now, we have other options that Consumers doesn't have because Consumers is a large-scale operation. The private citizen in this country has the option of going to small-scale operations, wind power, solar power, all the coming technologies . . . . And also burning wood for heat which is another small-scale thing.[17]

The exchange that ensued revealed ideological differences between the promoters and the opponents of the WFPP.

> *Proponent*: We're not adverse to that [that is, small-scale use of wood for home heating].

> *Opponent*: One place is making a decision between your large-scale operation and . . . the option of trying some small-scale thing.

> *Proponent*: But [there are] problems with all those, and this is part of my job with Consumers . . . is looking at various options. [He then recited some problems with the technical feasibility of small-scale alternatives using wind power as an example].

> *Opponent*: What you're finding out is that things like solar and wind technology are small-scale technologies and they really don't pay off for something like Consumers.

> *Proponent*: No, they would pay off for Consumers if they'd pay off for the small individual. I don't think there's anything that says we couldn't form a subsidiary that didn't promote solar energy and we would do that, I would think, if it were a feasible option. We're in the energy business.[18]

Opposition to WFPPs became regional in scope pursuant to the Hersey controversy. It was explicitly in tune with the supranational movement against large-scale energy technologies. CRUF sought to protect the forests of Michigan in the public interest, arguing that replacing labor-intensive with energy- and capital-intensive technology would curtail small-scale logging and firewood use.[19] By means of such arguments it was able to mobilize local interests, not the least of which was the potential threat to local firewood supplies. Implicit in this threat was the effect of scale in use of the resource. Wood stoves were preferred to WFPPs on ideological grounds.

Plant promoters, on the other hand, touted the interest of the state in meeting future energy needs from indigenous sources. Their ideology was consistent with conventional large-scale development of the resource. The local community would, of course, be called upon to serve the needs of the larger communities constituted by the state and the nation. Nevertheless, they argued, local interests could be well served by the additional tax base and employment opportunities created by the demonstrational WFPP. The overlapping ideologies of economic growth and hard path energy development undoubtedly appealed to many in the Hersey community, but concern for environmental quality provided fertile ground for the ideology of soft path development.

The opposition in the Hersey case may have been motivated by a soft path ideology, but the outcome depended more on the threat to local autonomy. Like other communities Hersey was reluctant to bear the costs of a project that would benefit other geographic areas. Hersey residents did not want to sacrifice their environmental values to the state's need for indigenously produced energy. Evidence for this lies in the changing priority of issues in the controversy. Analysis of issues in the Hersey controversy (see table 5) showed that local autonomy changed rank from sixth in the first phase to second in the last phase of the dispute. This shift is evident in the increasing use of locally oriented arguments by opponents to defeat arguments for the common good made by project promoters. Although arguments for local autonomy were more frequent toward the end, it seems likely that they were inspired by other issues. A closer look at the changing mix of issues suggests that the local autonomy as issue was instrumental.[20]

Some of the values held by the opposition and the implicit role of local autonomy were lucidly stated by one of the leaders of CRUF:

It's the to-hell-with Hersey attitude that bothers me. The plant, if constructed, will affect people's lifestyles. The people here know that the area is depressed and jobs are hard to find. But people choose to live here anyway. The pace is slow. The air and water are clean. We're not willing to take a risk with this woodburning plant. I'm of the mind that individuals must act in order to protect their rights.[21]

This remark, more than any other in the debate, sums up the nature of the controversy and the fundamental basis for social action. The argument concisely states the important values in the dispute.

This study of the defeat of the Hersey WFPP suggests that local political response to large-scale energy resource development is intimately connected with social change in nonmetropolitan communities. Newcomers brought by the population migration turnaround become integral members of local groups opposed to controversial technologies. Significant numbers of newcomers arrive in such communities with an ideology reflecting urban environmental values and the specialized knowledge and skills for political action. In this case they engaged both cosmopolitan (for example, the two-paths energy development debate) and local control (NIMBY) ideologies to mount an effective opposition.

8

The Politics of Expertise in the Hersey Controversy

A Politics of "Facts"

The siting of power plants epitomizes one aspect of public debate in techncial controversies—the conflict over who really has the facts. Advocates of such projects often contend that "if the people learn the facts" more support for the proposal will follow.[1] Getting out the facts seldom has the intended effect. Take, for example, the claim by one of the proponents of the Hersey project about WFPP development in general, that is, "we need to get the facts to the people who make the policies and control the money."[2] As it happens, this proponent headed the branch of Morbark Industries that promoted WFPPs. His viewpoint, though quite typical, is usually wrong. Similarly, the claim that getting out the facts will increase public support does not stand up to scrutiny. In controversies being highly informed means an increased awareness of the various positions in the conflict, a condition that arouses uncertainty and enhances consideration of the value issues that inexorably supersede the factual issues.[3] The acrimonious debate at the end of the Hersey controversy demonstrated a similar penchant for the "facts," when in reality values were the uppermost concern.

Acting upon their ideologies, both proponents and opponents made use of experts to advance their cause. As is usually the case in technical controversies, expertise was the prerequisite for participating in the debate. Experts confidently ushered in by promoters to explain the project—to get the facts to the public—were soon called upon to answer the challenge of opposition experts. The technical documents of each, respectively the Daverman and Associates feasibility study[4] and the CRUF position paper,[5] were the touchstones of the debate. Despite the lopsided financial clout of the venture partners, a fact reflected in their slick, voluminous, and expensive feasibility study, the local community seemed equally prepared to consider opposing technical arguments, however humble in comparison to the Daverman feasibility study. This was in sharp contrast to the experience of opposition experts in their dealings with state agencies, in particular, the Michigan NRC. In the end local politics determined the outcome as it very often does in siting controversies. Opponent experts succeeded largely because they capitalized on local issues. Arguments couched in terms of local autonomy were crucial to their success.

For several months after the decision of the Hersey township board to provide a referendum on the waste burning permit, the *Osceola County Herald* served as a forum for a plethora of recriminations by both sides in the controversy. In the midst of this exchange the Hersey site was ruled out altogether by the utilities during the second week of May 1980. The fascinating aspect of the belated give-and-take was its thoughtful presentation of the technical issues in the controversy. Once again utility representatives emphasized the "facts" in support of the WFPP. They harangued that theirs were the only legitimate experts. With their experts to establish the facts, the promoters wondered how the Hersey community could reasonably oppose the project. To quote the Hersey WFPP project manager:

> ... If the re-evaluation process is an extended one and if Hersey Township voters eventually face the RDF issue at the ballot box, I can only hope that they LOOK CLOSELY AT BOTH SIDES OF THE ISSUE—the factual side presented by Wolverine and Consumer Power Company versus the imagined-emotional side presented by the Committee for the Rational Use of Our Forests—before casting their vote.[6]

In the same article, the project manager assailed the opposition as undemocratic, accusing CRUF of "disrupting an orderly democratic process designed to exchange factual information upon which reasoned and intelligent decision making should be based." With the expertise at their disposal utility officials assumed in these exchanges that facts were what the controversy was really all about and that they had cornered the market on facts.

The Role of Experts and Expertise

The role of the experts was laid bare by this preoccupation with facts. Notwithstanding their moral fervor, even opponents of the Hersey WFPP focused presumptively on "the facts." Out of necessity they joined the debate with technical information and expertise designed to dispute the studies and experts of the promoters. Their corresponding interest in "the facts" is lucidly described in this response to the project manager's statements:

> It strikes me as almost comical that Consumers Power's Mr. Konchar accuses CRUF of not presenting FACTS in our fight against the wood-chip plant.... Members of CRUF spent six months intensively researching the concepts of whole-tree chipping and wood-fired electricity. We published the results of this research in a 22-page impact statement entitled "Wood Energy in Michigan: An Analysis of Impacts and Alternatives to the Proposed Generating Plant at Hersey, Michigan."[7] Both Consumers Power and the DNR have tried to refute the facts presented in this paper, and have been forced to resort to political maneuvering in Lansing and "arm-twisting" of local officials to accomplish their ends. Consumers Power has refused to consider the facts presented in this paper.[8]

This statement also points up the deliberate use of expertise by the opposition group in responding to the technical challenge of the promoters' feasibility study. It implies that technical information (read "expertise") was the medium of the controversy. The opposition group hoped to rise to the challenge of the well-paid experts engaged by the promoters of the project.

The basis for the controversy was further summed up by CRUF leader Marco Menezes in an interview with the *Osceola County Herald*:

> Their [project promoters] attitude was extremely condescending, like, "we're the experts, and we know what's best for you," and that just didn't sit well. Finally, a group of us decided to put our heads together and get some answers of our own.[9]

The development of the opposition was elementary by comparison with the elaborate and expensive organization of the project and the research effort of its consultants. A small group of Hersey residents banded together and formed the Committee for the Rational Use of Our Forests to provide the mounting opposition with an organization to rally around. Suspecting the impartiality of foresters, biologists, and other specialists connected with the venture partners, CRUF members then embarked on an extensive research project of their own.

> It didn't take long for CRUF to develop its own panel of experts ... with credentials every bit as impressive as those of the proponent specialists.

> Our people found glaring inconsistencies in their feasibility study ... and, for the most part, professed diametrically opposite conclusions to those of the project experts. These differences weren't just on minor issues either—we weren't interested in splitting hairs or debating fine points—and the inconsistencies seemed to arise on far more than an infrequent basis.[10]

This retrospective reads almost like a guide to opposing large-scale projects. The apparent strategy of the opponents was to counter the expertise of the project promoters. Events proved this to be far more effective than the corresponding strategy of the proponents meant to depoliticize the social and political issues with the assistance of their experts. The nature and method of the opposition experts is explicit in this statement.

The outcome demonstrates that credentials alone are not the path to legitimacy. An example of the legitimacy CRUF was accorded by the Hersey community is a letter by one resident who jumped to CRUF's defense: "were it not for the research efforts of CRUF, this project would have sailed into Hersey on empty promises based on misleading information."[11] Once CRUF acquired this legitimacy, it could debate with the promoters of the WFPP on equal terms. This effectively shifted the decision away from the technical experts employed by the promoters to define political decisions as technical.

The experts for the proponents had a high profile from the outset of the controversy. They were well-trained technicians and scientists on the staff of the venture partners or they were hired as consultants. In either case, they were well remunerated for the job they were expected to do. On the other hand, the experts for the opponents were unpaid volunteers largely from the local community. In most cases they possessed modest levels of technical training and experience by comparison with proponent experts. They did not measure up to the credentials of the experts on the other side. Nevertheless, the expertise of the opposition group proved adequate in challenging the legitimacy of the experts employed by the WFPP promoters.

Disagreement among experts over technical matters invariably directs public attention to social and political values—that is the net effect of technical controversy. In matters of values technical expertise is not a prerequisite. The outcome of the Hersey controversy aptly demonstrates that, although financial and political resources can be used to hire and legitimate expertise, this does not guarantee that the decision will remain closed or that local communities will stand mute when their values are threatened.

Some Propositions on Experts and Expertise in Technical Controversies

This is a revelatory case study. As such it is an opportunity to form generalizations for further study. A number of other studies and statements on experts and expertise have been cited in the foregoing chapters. The process of drawing out of generalizations was seen from the beginning as the means to put the results of the Hersey case into the context of this work. Accordingly, this section distills a set of propositions from the Hersey experience for summarizing the case and for directing further social study of technical controversy.

Dorothy Nelkin's comparative study of a proposed nuclear power plant in Ithaca, New York and a proposed airport expansion in Boston are the source of the initial propositions.[12] This is a classic study in the sociology of expertise. The propositions she derived from this study have proven their resilience since the article was published in 1975. Unless otherwise indicated, quotes accompanying explanations are taken from this article.

- *Developers seek expertise to legitimize their plans and they use their command of technical knowledge to justify their autonomy.* Developers generally assume that the expertise under their control obviates outside interference even from the public affected by expert decisions. A similar strategy is employed by public agencies. Experts, both public and private, adopt a "leave it to us" attitude.[13] This is in keeping with the scale of activities in modern society. Large-scale systems require massive information resources and call for an army of technical experts to manage the flow of information. Yet this expertise is often strategically used in ways that affect the public interest.[14]

- *While expert advice can help to clarify technical constraints, it is also likely to increase conflict.* Citizens' groups seek their own expertise to neutralize the impact of experts backing large-scale projects. The resulting disagreement between experts calls for a political solution.[15]

- *The extent to which technical advice is accepted depends less on its validity and the competence of the expert, than on the extent to which it reinforces existing positions [values].* This proposition unveils the incipient hypocrisy of the expert's role. It describes how, in fact, experts and expert knowledge affect the controversy. In technical disputes the efficacy of expertise is entirely different than that for "truth" by means of the scientific method. The measure of expertise is not scientific. Rather, it is political and social.[16] Technical experts are proxies for the conflicting sides in a controversy.[17]

- *Those opposing a decision need not muster equal evidence.* "... it is [often] sufficient to raise questions that will undermine the expertise of a developer whose power and legitimacy rests on his monopoly of knowledge or claims of special competence." Opponents take advantage of the asymmetry faced by developers in such conflicts— they must respond to challenges at every level of government or lose.[18]

- *Conflict among experts reduces their political impact.* Experts have influence because of the myth of their infallibility. Public trust is eroded when experts present conflicting information. Such conflict "highlights their fallibility, demystifies their special expertise, and calls attention to non-technical and political assumptions [values] that influence technical advice."

- *The role of experts appears to be similar regardless of whether they are "hard" or "soft" scientists.* The role of lawyers, economists, engineers, and scientists in the Hersey controversy was similar to that observed in Nelkin's comparative study. The expert's role appears to be generic. To the extent that this true, the continuum in figure 1 may be a useful model for describing the role of the expert.

The following propositions further refine and extend Nelkin's propositions based on the present study. A brief explanation is appended to each proposition.

- *Technical controversies generally focus on factual claims even though the underlying concerns are political or ideological values.* This proposition makes explicit the arena in which the dispute takes place. Experts and counterexperts haggle over the facts on the presumption that they are the "data" for crucial decisions. Beneath this veneer are the value concerns that gave rise to the controversy, the unheralded ingredient of ostensibly technical decisions.[19]

- *Although it can serve to clarify technical issues, technical expertise in effect stimulates conflict predicated upon existing values.* The ideal outcome for project promoters is acceptance of their experts as legitimate, their expert claims as true, and, accordingly, unqualified acceptance of the proposed project. Historically this strategy has been successful, but it became exceptional beginning in the 1960s. Use of experts to clarify technical issues only creates further conflict because, as implied by the foregoing propositions, the facts are not the essence of technical controversy. Regardless of how straightforward the facts are, factual arguments educe latent value concerns. The reverse also is evident in controversies. Value issues lead to increased debate about technical issues. This proposition formally states that the very existence of technical expertise is a potential challenge to group values. However well the experts lay out the facts, the act of doing so for a particular interest sets the stage for controversy.

- *Notwithstanding a general focus on factual matters, a rise in controversy tends to shift the focus of the dispute from technical to value issues.* The dispute evolves, according to this proposition. The project is proposed and experts are called upon to reinforce the idea. Contingent on its values, the opposition brings forward its own experts to examine the facts, dispute them, and reach a different conclusion. Over time latent value issues surface for all sides in the controversy. The facts continue to be contested; the debate seeks to remain technical. However, values become increasingly apparent in the controversy. It takes on the character of a political debate. Threats to community autonomy, prospective degradation of the environment, loss to the tax base if the project is rejected, and like issues emerge during technical disputes. These issues more or less displace an evaluation of the facts per se. Both sides try to influence public opinion, appealing to values they believe the public holds dear. The technical debate may stick closely to factual matters in legal or administrative forums, but for the debate as a whole values gain the upper hand.

- *While technology is, in general, disruptive of older values, social changes accompanying technological progress, including the environmental movement and the population migration turnaround, reinforce the tendency to resist the disruptive effects of technology.* Currents in social change have their effect on technical controversy. Values have changed in the countryside as well as the city. Members of the community increasingly rise to the challenge of providing the technical expertise to fight controversial development. Social change has made the strategy of using technical expertise to depoliticize issues less useful to promoters of large-scale technologies. Balancing this tendency with a countervailing expertise is an important social change for democratic societies. Citizens are demanding a say in technical decisions and are achieving it one way or another.

- *In technical controversies local opposition groups advocating a cosmopolitan ideology utilize local interests to resist the plans of project promoters.* New ideologies about the environment and energy use are widely appreciated and often applied to the fight against developers. Leaders of the opposition were quick to learn that the NIMBY tendency could be used to halt unwanted developments. At the same time, changing values find many longtime residents of rural communities in tune with the ideologies opponents seek to effect. In any event, old-timers and newcomers readily cooperate when threatened by outsiders who disregard cherished community values.

These propositions constitute a summary of the role and impact of expertise as demonstrated by wood energy development controversies. They suggest that, despite claims to objectivity by experts, unexpressed values very often motivate them. The use of expertise in conflicts mirrors the contending positions and represents a strategy for influencing the decision-making process. When experts who hold different values counter establishment experts, value concerns begin to override questions of technical alternatives.

The Strategies of Expertise Revisited

The Hersey controversy is especially fascinating because it took place in a remote rural community. In the not too distant past you would not have expected to find a viable protest is such small towns so far removed from urban centers. The values of longtime residents in nonmetropolitan communities have generally been at odds with such protests. However, the "oldtimers" of Hersey possessed other values—emergent environmental values—that made them susceptible to mobilization by newcomers. A myriad of recent rural protests demonstrate that newcomers and oldtimers can cooperate for their mutual benefit. In the Hersey case valued environmental amenities proved attractive to disenchanted urbanites from areas to the south. The clean environment and scenic beauty of the Hersey area, perhaps in conjunction with absence of the disamenities of · urban life, brought a sufficient number of environmentally conscious newcomers to the area for a community to form. They came to Hersey with diverse values and perspectives. This new pluralism was an important change, but so too was the newcomers' tenacious interest in environmental values. Similar values were present but latent among oldtimers. The winds of change among the opposition leadership showed in their character. They were young, intelligent, and well-educated. Concomitantly, they possessed an ideological commitment to preserving the values that had attracted them to Hersey. These penchants, combined with their abilities to organize and to utilize counterexpertise, lead forthwith to the defeat of the Hersey WFPP project.

The Hersey case demonstrates that opposition groups in nonmetropolitan communities can defeat presumably benign alternative energy projects by emphasizing threats to environmental values and to the autonomy of the local community. Newcomers and oldtimers readily cooperate to protect amenities that both value. Conflict between developmental and environmental values evident in such controversies soon manifests the strategies of expertise. The complexity of technical debate and its attendant politics belies this tendency. Unless and until opponents acquire a strategy involving technical expertise, developers have it their own way. But even in small, remote communities, well-heeled developers can go down in defeat if opponents are poised to take advantage of this strategy. Wood energy controversies reveal how effective the strategies of expertise can be in affecting technological decisions despite the widely unequal resources of proponents and opponents.

Strategies of expertise do not arise in a vacuum. They are directly traceable to social values. In addition to the technical alternatives delegated to experts, choices in technical controversies necessarily involve competing values. Technical knowledge becomes instrumental for divergent interests to establish their claims. The criteria applied to interpretation of technical information often depend upon the values of the interpreter. Experts disagree on the facts, to be sure, but also on the basis of values. The resulting drama of values and facts is played out on a stage of technical debate in the public media and in administrative and judicial forums. Conflicting values and disagreements among experts inevitably force a political solution. Decision makers beholden to an electorate tend to fall back on ideology, interest group pressure, and the like for making decisions. To do otherwise is to risk losing office. The strategies of expertise thus push the technical debate into the political arena where values can be directly addressed.

The strategic use of experts and expertise in democratic societies has evolved in response to the bureaucratic tendency to narrow the decision-making process to technical issues to be decided by carefully managed experts. The strategies of expertise—expert versus expert in political disputes—serve to open up the decision-making process in an age of accelerating technological development. This would appear to be a natural response to the growth of technical knowledge and its political uses.

9

Understanding Controversy: The Case of Wood Energy Development

Strategic conflict between experts and counterexperts is the essence of this study. But there are wider implications of wood energy controversies for the social study of science and technology. Conclusions and implications for technical controversies in general merit discussion. Given the fit of the wood energy controversies with other cases, this study may be instrumental in forming generalization about technical controversies.

Controversy is a complex phenomenon which confounds attempts to simplify it to a small set of quantifiable variables. Facts and values are the elements of controversy. Ideology defines values in terms that facilitate action upon them. The stages and style of technical debate are strikingly similar among technical controversies, whether they concern WFPPs, radioactive waste facilities, or power lines. This chapter goes beyond the strategies of expertise to fuller examination of the dynamics of controversy.

Values and Facts in Controversies

The institutional use of expertise entails defining social and political values as technical problems. In this form they can conveniently be relegated to expert judgment. But of course, experts are paid and they know what is required to keep the paychecks coming. In this context the existence of "a persistent set of value choices even in the most objective and expert of governmental functions"[1] cannot be obscured from the opponents of large-scale wood energy technologies. Volunteer experts face off against experts in government and industry, presumably to debate the facts. The reality of controversy is far different. It has little to do with disagreements of fact. Values are fundamental to the conflict even if they are not explicitly addressed.

208

Factual uncertainties are quickly exploited in controversies. They give rise to diverse and value-laden interpretations. Technical questions become controversial because of the unavoidable ambiguity between the technical and the normative, that is between facts and values. Choices between different political and social values are the nexus public concern. Resolving technical questions one way or another quickly moves to the background in a controversy, particularly in environmental controversies where serious risks and hazards threaten the local community.

Controversy encourages a political solution. Decision makers find it convenient to embrace community values or simply to avoid the decision. Technical advice is soon measured by how well it reinforces the values of one side or the other. The asymmetry of financial resources for developers is counterbalanced by the advantageous position in which opponents find themselves—that is, it is often suffcent to raise questions that dispute expert conclusions. The resulting uncertainty and confusion reinforces a political approach. An accounting of credentials and evidence seldom takes place, even if some experts may brandish a Ph.D. to cast doubt on the credibility of their counterparts.

Confounding this pattern is the fact that values change. New values have emerged to dominate the environmental era. The values the gave rise to the population migration turnaround derive from general changes in the public perception of the environment. These changing values have come into conflict with traditional developmental values.[2]

Siting controversies are typical. Concern about the quality of life in the community is the wellspring of opposition. Yet the debate focuses on technical matters. To be technically effective opponents must seek to manipulate knowledge, emphasizing areas of uncertainty that are open to conflicting scientific interpretation. In this way citizen opposition groups can seek to justify their political and economic views. Political values and scientific facts consequently become difficult to distinguish in controversies.[3]

One practitioner eloquently describes the essence of facts and values in technical controversies as follows:

By seeking to maintain a strict separation between facts and values, especially on issues of great public controversy, government has facilitated the task of technical experts, but often at the expense of its own credibility. This is because, rightly or wrongly, the public is usually less interested in the facts of a given controversy—especially when the facts conflict—than it is with the choice between different political and social values.[4]

Although technical controversies are not just about technical alternatives but more importantly about competing sets of values, experts continue to be marshalled to impartially render decisions strictly on the basis of the "facts." However, the experts are not neutral, as the last chapter demonstrates; values held by experts influence their assessment of the facts. That the role of expert persists attests to the high premium our society places on efficiency. Experts define political problems as technical problems. It goes with the job.[5]

The analysis of facts, values, and technical expertise in the Hersey controversy summarized in chapter 7 supports the notion that latent values are brought out by controversy. Experts on both sides led the debate. The experience of other wood energy controversies is only suggestive on this point but they appear to follow the Hersey pattern. Controversy seems to encourage a trend toward more explicit treatment of values. This is turn shifts the focus of the debate away from the facts. However, the discussion does not stray far from the technical issues because this is the common ground on which the debate takes place.

When the adversarial approach takes its inevitable toll, the victor and the vanquished are likely to squabble over who really possessed the facts—as if this were the fundamental basis for the conflict.[6] The period of recriminations in the Hersey controversy noted in chapter 8 seems to be typical of protracted technological controversies where the outcome is unequivocally for one side or the other.

If this study of wood energy controversy does no more than clarify the dynamics of values and facts, then its purpose will have been well served. Controversy is a social and political phenomenon that should never be confused with the technical issues under dispute. A minimally adequate account of controversy must consider the values at issue, the ideologies of the partisans, and the role of experts and expertise. To maintain that the facts are the only or the most important data for decision making constitutes a rationalistic bias that precludes understanding of technological controversies.

The Ideological Basis of Controversy

Ideology is a strategic tool used by activists to recruit and mobilize followers. It describes social reality and prescribes social goals. To some, technical controversies may seem like knee-jerk reactions to the imposition of a burden on the community. However, the rationale for the opposition usually runs deeper than simply labeling the controversy as a NIMBY would suggest. To be sure, communities oppose hazardous and risky technologies because of NIMBY. Deep-seated value changes reflected in the ideologies of opposition leaders provide a more subtle but more compelling reason why large-scale technologies are being challenged at every turn.

The ideologies of developers and opponents alike are the essential mechanism of controversy, giving rise to the strategic use of experts and expertise to accomplish their respective goals. The means by which ideology is translated into action has been traced in this study from values to the strategies of expertise. In the process, the dynamics of controversy has been illuminated, and with it, the involvement of citizens in technological change.

Stages and Style of the Technical Debate

The wood energy controversies in this study fit several models proposed for local disputes. Three are discussed here—a general model (table 18), a model for nuclear power plant siting disputes (table 19), and the pattern of technical debate observed for controversial technologies of speed and power.[7] The similarity between the Hersey case and the third description is almost uncanny:

> The developers . . . contracted for detailed plans of the construction of their proposed facility. As they applied for the necessary permits, affected groups tried to influence the decision. The developer in each case argued that plans, based on their consultants' predictions of future demands and technical imperatives concerning the location and design of the facility, were definitive.[8]

TABLE 18
Stages Common to Local Disputes[9]

I.   Initial plans to introduce the innovation.

II.   Proponents submit a proposal to appropriate governmental authorities for approval; the first opposition appears.

III.   The opposition gains broad public support; active civic group forms; public meetings, demonstrations, and election-like campaigns characterize the conflict; the media become attentive; disagreement occurs among experts.

IV.   A crucial decision is reached by some governmental authority, or else the proposal is withdrawn. However, the issue is not necessarily settled. The defeated party may appeal against the decision or raise the issue again later on.

Source: Mazur, "Opposition to Technological Innovation." He relies in part on Jopling et al., "Forecasting Public Resistance to Technology" (see table 19).

TABLE 19

**Jopling Model for Stages of Resistance to
Nuclear Power Plant Siting**

STAGE:	1	2	3	4
DESCRIPTION:	Public Disclosure	Expert Inquiry	Information Distribution	Citizens Organize
TIME:	Months	Days	Months	Month or two

STAGE:	5	6	7
DESCRIPTION:	Technical Disagreements	Uncompromising Conflict	Legal Confrontation
TIME:	Months	Year or more	Unlimited

Source: Jopling et al., "Forecasting Public Resistance to Technology."

The Daverman and Associates feasibility study was the critical take off stage of the Hersey dispute, just as the five-volume technical report had been in the Lake Cayuga controversy and the environmental impact statement in the Boston airport controversy.[10] Informational meetings, in each case rife with dissent, played an important role in the latter as well as the Hersey case. Furthermore, the Hersey dispute, like Lake Cayuga, involved a critique of the consultants' technical report that contributed to the polarization of community views.

In general, the wood energy controversies studied fit the models in tables 18 and 19. The stage of disagreement among experts (table 18, stage III; table 19, stage 5) marks the turning point in technical controversies. What might be termed the advocates stage (table 18, stage IV; table 19, stage 7)—the period of adjudication—in the Hersey case involved the passage and amendment of a local toxic and hazardous waste ordinance and at Indian River resulted in an injunction. The environmental lawyer retained by CRUF to draft the ordinance, the Hersey Township attorney, and legal counsel for Consumers Power, were all involved in the compromise amendment. This legal confrontation in the period preceding the climax of the controversy was a short-lived solution to the conflict at Hersey.[11] Entrenched community mistrust of the intentions of project promoters soon precipitated outright withdrawal of the proposal (table 18, stage IV; table 19, state 7).

The style of the technical debate in the Hersey controversy and the others reviewed here is very similar to that in the cases observed by Nelkin, that is, "considerable rhetorical license, with many insinuations concerning the competence and biases of the [experts] involved."[12] Some of the rhetoric was used for the purposes of illustration in chapter 8. Two especially poignant exchanges in the *Osceola County Herald* took place during and after the conflict over the application for a permit to experimentally burn RDF in the WFPP. The debate between columnist Walt Grysko and Wolverine Electric president John Keen beginning in November 1979 was quite vituperative, verging on personal attack. The exchanges between CRUF members along with some Hersey area residents and the project managers from the utilities were more low keyed and tied to the issues. The *Osceola County Herald* served as the forum for this debate. Of the seventeen letters to the editor of the *Osceola County Herald* throughout the controversy, twelve were published during this postmortem phase.[13]

The media debate concentrated on who had the facts. The divergence of opinion on the facts of the matter squares with Nelkin's finding that each party in the disputes used different criteria to collect and interpret technical data. Her conclusion on the style of technical debates is appropriate here:

> Both disputes [Lake Cayuga, Boston airport] necessarily dealt with a great number of genuine uncertainties that allowed divergent predictions from available data. The opposing experts emphasized these uncertainties; but in any case, the substance of the technical arguments had little to do with the subsequent political activity.[14]

The style of these wood energy controversies was also relatively sedate compared to other recent energy-related protests. The fight over power lines in Minnesota came to a head in 1980, simultaneously with the Hersey controversy. While Hersey opponents methodically debated the merits of WFPPs in every available forum, protesters in Minnesota were disabling power lines with high powered rifles and toppling one-hundred-fifty-foot towers by loosening their bolts.[15] The specter of farmers supporting and actively participating in such protests underlines changing values relevant to technological developments in nonmetropolitan areas today. It also serves to point out the different forms conflict can take when technical controversy remains unresolved.

Experts and Expertise in Wood Energy Controversies

The use of expertise as a tool both to support and oppose projects is an unmistakable social change ushered in by the environmental era. The role of experts has become more egalitarian. Opposition groups have learned to cultivate their own experts to neutralize the technical expertise of project promoters. Voluntary experts will undoubtedly continue to carve out a niche in technical controversies.

The importance of the values held by an increasingly diverse rural polity has been emphasized by this study. Given the role of experts and the complexity of the energy problem, the gap between public acceptance of small-scale and large-scale energy technologies may not be narrowed by continued resort to expertise. Whatever policy ultimately guides future energy development, the practice of seeking to depoliticize issues with the aid of experts will undoubtedly be reevaluated as citizen opposition groups continue to stymie large-scale projects.

The emergence of an alternative expertise that could significantly influence decision making is an unheralded phenomenon of the 1970s. The social role and political impact of experts in general has not been accorded the scholarly attention it deserves. Voluntary experts can be expected to continue to play an important role in technical decisions given trends in the decentralization of controversies and decision making.

The present discussion should alert social science scholars and practitioners about the emergence of voluntary experts. To quote Nichols: "What is required is a clearer understanding and articulation of the appropriate role and limits of technical expertise in processes of democratic decision making."[16] More and better case studies should prove fruitful both for the advance of knowledge about technical controversies and the enhancement of democratic means for rendering decisions about the burgeoning class of technologies that are a mixed blessing.

10

The Challenges of Technocratic
and Democratic Decision Making

The clash of values and interests, the tension between what is and
what some groups or individuals feel ought to be, the conflict
between vested interest groups and new strata demanding their
share of wealth, power, and status are all productive of social
vitality.[1]

Within the context of a constitution designed to protect the
people from government and to trace all power back to the
governed, the people are the experts. They must be consulted
regarding value choices.[2]

A variety of general conclusions might be reached based on this
comparative multidisciplinary study of wood energy development.
Some would be of greater interest to research scholars—others to
natural resource policymakers and practitioners. However, together
these conclusions would provide just a partial account of controversy. Wood energy development is also a particular kind of
environmental problem that in some respects may not be applicable
to other environmental controversies.

With these limitations in mind, I have selected conclusions and
implications in three fundamentally important areas. They examine
respectively: how to interpret environmental controversy; how to
understand the response of local communities faced with controversial developments; and the role of technical debate as it is and as
it might be in a democratic society. This chapter extends the
foregoing analysis and discussion of wood energy development,
discussing implications for a wider set of technical controversies. The
conclusions are stated at the outset as a set of hypotheses for making
sense of controversies. They constitute a brief summary of the
chapter.

- Technical controversies entail elaborate debate about factual issues,
  masking the underlying values that are the essential basis of conflict.

215

- Local communities, even in remote nonmetropolitan areas, increasingly resist sacrificing local autonomy to projects that threaten the quality of their environment.
- Citizen groups have learned that expertise is a political resource, and have become sophisticated in utilizing voluntary experts to oppose controversial developments.

Each hypothesis is treated in turn.

Interpreting Environmental Controversies

Environmental controversies are not about facts alone. It is a mistake to think that the issues are largely technical and therefore best relegated to the experts. Public distrust of government and of science has increased in proportion to the use of expertise as a political resource. Accordingly, environmental controversies must be fundamentally viewed as expressions of values as well as facts. Any explanation confined to the factual issues will miss the essential basis of controversy.

As obvious as this may seem, many persist in the view that controversies are misunderstandings about the scientific facts. When faced with controversy, it is all too common for government bureaucrats and "the experts" to call for public education. Proposals like the science court and, to some extent, environmental conflict mediation, mistakenly take resolution of factual disagreements as resolution of the conflict. Environmental controversies must be seen for what they are—political conflict based on values which is expressed ostensibly as technical debate.

This political conflict is not about technical alternatives; it is about values. The facts become less crucial in technical controversies, particularly when experts disagree about them. In such circumstances, the public becomes keenly interested in the values.[3]

The findings of this study reinforce similar results from studies of other environmental conflicts. The WFPP controversies reviewed here had many common features and followed the general pattern of behavior for power plant siting controversies. The ideologies and values of the affected communities were in a genuine sense more relevant than facts about the technology and natural resource per se. Values proved essential to explaining the conflicts, particularly given the remote nonmetropolitan locations of many of the proposed WFPP sites.

The values held by an increasingly diverse rural polity, combined with a penchant for involvement in political affairs by newcomers to these areas, have created the conditions for opposition wherever local values are threatened. When opportunities to give voice to their values are not provided by the political process, WFPP opponents use the politics of protest. This should be understood as an attempt at democracy by default. Decision makers may find that providing opportunities for these values to be expressed will reduce incidences of conflict and lead to a more orderly development of wood energy.

Energy development, especially the siting of large-scale facilities, can be expected to rise on the agenda of policy makers. Fossil fuels are being depleted and at a growing rate. In the United States imports of foreign oil have increased to an all-time high, yet political developments in the Middle East could disrupt or attenuate the supply at any time. The Hersey case demonstrates once again that the level of public concern about environmental quality and especially the hazards of power plants remains high. This concern extends even to large-scale renewable energy resource development in which the environmental risks are regarded as low. It is important to understand that the individual consumer assumes his or her own risks in choosing small-scale, decentralized technologies. The social choice for large-scale energy development is qualitatively different. This choice requires opponents to acquiesce to risks they do not wish to assume. In the wood energy cases studied here the public did not object to the small-scale uses to which the resource had been put. Indeed, the threat to these uses was a factor motivating opposition to WFPPs. Energy policy that recognizes the growing interest in small-scale energy technologies would seem to be both more responsive and less conflictual than exclusive development of large-scale, centralized power plants.

Natural resource scholars and practitioners should seriously consider the issue of scale as it relates to public perception of resource policy and management. The Hersey case suggests that the two-paths ideology of energy development may have broader application. The issues in the dispute transcended the production and consumption of energy. One of the persistent arguments was that local loggers and their small-scale operations would be displaced by whole-tree harvesting and the widespread use of wood chipping machines. Natural resource conflicts involving environmental issues may be better understood if they are looked at in terms of their associated ideologies. The division between small-scale and large-scale development is apparent in other natural resource controversies as well, suggesting that an examination of their ideological basis may be fruitful.

The scale of government as well as technology is an issue for technical controversies. Technological scale is the factual, but value-laden issue addressed by the first hypothesis. The second hypothesis deals with local autonomy, which in turn is an issue of governmental scale.

Understanding Local Autonomy

Local government is increasingly the focal point of the ideology of local control as citizens are frustrated in attempts to effect change at higher levels of government. Centralization of decision making has made the locus of change more and more remote from the citizenry.[4] Decisions by big government and big business that affect the values of citizens are not likely to be redressed in the state capitol or the corporate boardroom. Citizens may have no voice there. Consequently, the best chance for citizen groups to be heard is in the community where they live. This notion has not been lost on opponents of large-scale technologies. Local autonomy—the privilege of deciding what happens in your community—has increasingly been asserted to reverse decisions made from outside the community.

Along with a proper interpretation of environmental conflict, this is an idea that decisionmakers should take to heart. Federal or state policy may not make the grade in communities like Hersey, Michigan, or Quincy, California. These decision makers should realize that failure to address genuine public concerns may preclude realization of a policy. Better communication among levels of government is one possible answer. Perhaps the best answer to this problem is finding better methods of including public values in the decision making process. Where the values of federal and state bureaucracies stand in opposition to local community values, then perhaps conflict is inevitable. When this happens it should not be seen as an aberration but as the democratic process at work.

Wood-electric power development demonstrates that a variety of social changes are tied to the response of local communities to technical controversy. Newcomers via the population migration turnaround figure significantly in public opposition. The specialized knowledge they bring to remote communities faced with proposals to develop WFPPs represents a human resource that can effectively be utilized to protect local autonomy in these decisions. There seems to be a clear connection in WFPP disputes between the turnaround, the widely recognized policy choice confronting our energy-intensive society, and the defeat of these controversial projects.

Renewable resources are by nature more likely to be found in nonmetropolitan locations. This study shows that it would be a mistake to take public support in such areas for granted, even in communities that need and want economic development. Nonmetropolitan siting of large-scale energy facilities is less likely to succeed today because of increased pluralism. The newcomers brought by the population migration turnaround possess values, interests, and skills for opposing controversial developments. Natural resource decision makers who ignore these facts do so at their peril.[5]

Citizen Participation versus the Politics of Expertise

Notwithstanding the desire of scientists and technicians to assert the neutrality of their enterprise, expertise functions as a tool both for supporting and opposing technological development. Opposition groups routinely find their own experts to dispute with the experts supporting development. Given the role of experts and expertise in our society and in view of the complexity of the energy problem, continued resort to expertise is likely to increase conflict predicated on public apprehension about the scale of energy development. Attempts at depoliticizing issues with the assistance of well-paid experts will very likely only foment public controversy. Whatever policy ultimately guides future energy development, the traditional practice of seeking to depoliticize issues by calling in the experts will be challenged as citizen opposition groups and their voluntary experts continue to successfully oppose large-scale development projects.

The role and impact of technical expertise typically observed in siting disputes was apparent in the controversies in this study, particularly the Hersey case. Opponents recognized that there are a number of environmental hazards and uncertainties in the development of wood-electric power. Emphasizing these uncertainties by bringing in opposing expertise has proved to be a fruitful means for defending local values in WFPP disputes.

Instead of doing battle with the assistance of experts, public officials and developers may come to realize that protest is a symptom that democracy is not working effectively. Public participation is a logical and much less costly alternative to conflict and delay. If I were ask to summarize the practical utility of this study in one brief statement, this would be it.

Environmental controversies surrrounding the introduction of large-scale technologies in general may not be fundamentally different from the cases studied here. If this is correct, these results can provide a framework for understanding a broader class of controversies. The foregoing conclusions regarding the interpretation of controversy, the significance of local autonomy, and citizen participation versus the politics of expertise would then be useful far beyond the problem of developing wood for electric power.

This study has proceeded in an orderly fashion for social research on an emergent, relatively unexplored phenomenon. It developed over the span of a decade. An original case was meticulously analyzed, in the process exploring a broad range of ideas to explain the behavior observed. This was followed by protracted stocktaking, refinement, and observation of additional cases. Each of the latter represented a parallel case ripe for comparison. In carefully reviewing each case for similarities and differences, it became clear that widely separated places had much in common with regard to technological controversy. The outcome was not always the same, just as the mixture of social factors thought to be significant varied from place to place. However, major similarities, especially in connection with the contribution of the population migration turnaround and the political use of technical expertise, supported the original case study. A number of generalizations have resulted from this endeavor. In turn these have spawned a number of implications for policy and public administration. Additional study will serve to test and refine the general case made here. In the meantime, the strategies of expertise can be understood as a fundamental aspect of controversies with important practical implications for the introduction of technologies in a democracy.

Technocracy versus Democracy

The challenges of technocratic and democratic decision making are the final order of business.[6] Wood energy development is perhaps just a minor area of concern in the scheme of things, but it does provide an opportunity to examine critically how we use experts and make decisions. Despite its humble beginnings, this information would seem to have wide application indeed.

The analysis of wood energy development highlights the use of expertise in technical controversies to surface value concerns for political debate. Controversy appears to serve as a social mechanism for reaching decisions. The choice may be up or down on the particular technological proposal, or it may be protracted conflict encumbering any related proposal. In any event, the strategies of expertise derive from the conflicting tendencies to depoliticize issues with technical expertise and to democratize issues with counterexpertise.

When proponents of technological proposals seek to define value issues as technical questions reserved for the judgment of experts, they disenfranchise citizens interested in these values. This challenge to democratic decision making has been met with the politics of protest guided by an understanding of the use of counterexpertise. Despite unequal resources, opponents very often succeed in defeating controversial proposals. The opposition process demonstrates:

> ... that citizen opposition groups are effective, at least in the short run, against the combined expertise, authority, and social legitimacy of the state and large corporations. The strategy used by these citizens, engaging in formal organization protest, is a general one available to members of any community in our society ... the tactics used are strongly influenced by recent legislation and court decisions supporting the concept of the right of citizens to a clean environment and proper administrative procedure.[7]

On the other hand, technologically sophisticated, highly centralized societies require an orderly process for rendering decisions about development. The persistence of organized efforts to neutralize environmentally hazardous proposals is a chronic problem for government. In recent years there have been a number of efforts to avoid controversy by increasing citizen involvement or to resolve conflict through environmental mediation. The results have not been good. Bureaucracies cling to the use of expertise as a political resource.[8] Citizens have seldom been allowed to penetrate this establishment to directly address issues of social and environmental values.

This study suggests that the politics of protest and counterexpertise are and will remain business as usual in wood energy development. The assessment of Michigan's wood energy policy shows that government bureaucracies are unenlightened by experience with controversy. Is it possible that a better solution to controversial development can be found? Any solution should be tempered by an understanding of the strategies of expertise and corresponding methods for clarifying value issues. The agencies of government can then involve the public in decisions that affect these values in ways that will minister to the concerns giving rise to controversy.

This is a tall order. The proponents of technology seem to

understand only one class of decisions—they tend to define decisions as technical. Business and government are all too willing to embrace this tendency because it reinforces their power. Notwithstanding the ubiquity of hearings on all manner of proposals, citizen participation is a rarity. One researcher put it this way:

> A great deal of lip service is paid to citizen participation in general in the United States, and participation in local community organizations is considered essential for a socially integrated society. Yet, as much as it is part of the ideology of pluralism, citizen participation is considered desirable only if the participants play by the rules. That is, accept the ideology that only those with the *formal* expertise and the *formal* responsibility can know what is good for the state and its citizens. Citizen groups are restricted to an "advisory" position in regard to the actions of monopolies and bureaucracies.[9]

The crucial issue is recognition of values as legitimate. This study has returned time and again to values in the technical debate about wood energy development. Values played a conspicuous role in the controversy. If there had not been a conflict of values, it is unlikely a controversy would have occurred. The significance of conflicting values points to the analysis of values as a worthwhile exercise. Practitioners in government should consider how to make values explicit and deal with them before controversy takes its toll. However, the study of values cuts against the grain of technocratic decision making. It is a direct challenge to experts and expertise as a political resource. To state the case more succinctly:

> It is not too much to say that one of the major political tasks of the present day is to cut through [the] weight of expertise, to lay bare the value implications of planning so that the public can understand the choices that are being suggested and make intelligent decisions about them.[10]

Laying bare values is also a challenge for academic researchers and scholars. Values may resist quantification, but that does not, and should not, remove them from scientific scrutiny. By the same token, researchers and scholars cannot really hope to avoid controversy. If the foregoing analysis of the role of experts and expertise is any guide, many academics will come to fill the shoes of the expert. They will then take part in ideological debates that transcend the norms and ideology of science. The student of controversy studies should be forgiven for seeing as myopic the view that science is completely separate from such nonscientific matters as values, ideology, and politics.

This is a problem not just for the institutions of science or government, but for the welfare of the nation as a whole. The president's Commission for a National Agenda for the Eighties articulates the case for better understanding and treatment of the overriding tension between technocratic and democratic decision making:

> The nation ... may be at a crucial point in its history; technological complexity and ability to regulate may no longer be in balance. If so, our decision-making processes need to be critically examined to determine whether they are adequate for future challenges. The national agenda for the next decade should include consideration of mechanisms consistent with a democratic society that permits public participation in decisions that potentially could affect the lives of many people—the uses of scientific and technical knowledge. During the decade of the eighties, our society needs to *examine critically the relationship between scientific expertise and democratic policy making* ... the management of increasing scientific and technological complexity pose a difficult challenge for public and private institutions, requiring flexibility and adjustment at all levels.... However, a nascent contradiction may exist between the requisites of science and the requisites of democracy. [Emphasis mine.][11]

Unless and until this crisis in decision making is resolved, the strategies of expertise will remain the driving force of controversy. Some may view this as a threat to the unity of science. Yet scientists and technicians disagree among themselves about scientific issues as well as about values. Perhaps the die of science should be recast to provide a more realistic account of the social roles of science and expertise. In any case, the progress of science is hardly threatened by the use of its practitioners and terminology to deal with underlying value issues—whereas the use of expertise to mask these issues is a genuine threat to democracy. The strategies of expertise, therefore, serve the purpose of bringing the value issues relevant to technical decisions into the political arena where they can be scrutinized and fairly judged. This conclusion suggests that the strategies of expertise portend neither a decline in the integrity of science nor the impending demise of democracy. Disagreements among experts and concomitant technical controversy should be viewed as healthy activities in a democratic society. The strategies of expertise have evolved out of a need to facilitate political reckoning of the value issues in technical decisions. The tenacity of the strategies of expertise and the fact that workable alternatives have not been found underline the fundamental contribution of this social mechanism.

The strategies of expertise serve a number of useful purposes. Among others things, they:

- Give citizens more direct opportunities to enter the process of decision making about large-scale technologies[12]
- Encourage officials to consider policy alternatives
- Increase the perceived range of choice; invigorate debate about new issues
- Shed light on aspects of decisions that might otherwise have been ignored
- Provide a grace period within which conventional political forces can reassert themselves
- In general, produce social vitality

Yet the adversarial approach implicit in the strategies of expertise is socially inefficient. Making the public a genuine partner in technical choices might help resolve the dilemma between technocratic and democratic decision making. Heightened citizen participation may be "a force waiting to explode in relation to a host of troublesome technological issues in the wings of the public policy arena," but the social mechanism is in place for dissipating its energy. Until the basis for conflict is appreciated and the solution found in satisfactory methods of citizen involvement, the strategies of expertise will continue to fulfill their purpose.[13]

Appendix 1: Hersey WFPP Controversy Chronology of Major Events, 1978–1980

Month	Promoters' Activities	Citizens' Activities	Relevant Hearings, Meetings, and Regulatory Activities
1978			
June	Joint venture to build a demonstrational WFPP announced by Consumers Power Company, Wolverine Electric Cooperative, and Morbark Industries.		
September	The three final round sites announced— Harlan, Whitehall, and Hersey.		Daverman and Associates holds environmental meeting at Big Rapids.
October		Concerned Hersey area residents attend hearing in Reed City.	Daverman and Associates conducts hearings at or near each of the three sites in the final round.
December	Big Rapids Industrial Development Commission endorses the Hersey site.	Hersey citizens ask that WFPP proposal be aired in Hersey. Citizens' study committee and Committee for Rational Use of Our Forests (CRUF) form in wake of Hersey meeting.	Public meeing held in Hersey Township Hall.
1979			
January	Visit Morbark by Burlington (VT) utility representative. Proposal for WFPP presented to State Public Service Commission and Michigan Department of Natural Resources (DNR). Daverman and Associates makes initial request to Michigan Natural Resources Commission (NRC) for commitment of wood from state forests. Sites narrowed to Hersey and Whitehall.	Research begins on CRUF position paper. CRUF holds several meetings and strategy sessions.	

Appendix 1—Continued

Month	Promoters' Activities	Citizens' Activities	Relevant Hearings, Meetings, and Regulatory Activities
February	Daverman and Associates feasibility study released; copies presented to Governor Milliken and Public Service Commission. Hersey site selected. Reply to CRUF's twelve questions in *Osceola County Herald*; reply also distributed at Hersey meeting.	CRUF produces a list of twelve questions about the WFPP. Prominent local columnist's first article on the WFPP—"Woodchip Plant Monster." Hersey mayor requests that proponents hold hearing in Hersey.	Public meeting convened in Hersey. Two hundred attend, most hostile to the proposal. Local state senator attends.
March	Morbark drops out of venture.	CRUF counters Daverman and Associates feasibility study with a position paper, debates proponents on TV, and holds public meeting to solicit more members.	Osceola County Commission passes a resolution in favor of the project.
April	"Pro: Money's a Key," slant on WFPP in *Osceola County Herald*.	"Con: Use is Inefficient," slant on WFPP in *Osceola County Herald*. CRUF presents position paper and petitions to Michigan NRC. Local senator does not show for presentation. Michigan DNR asked to produce better evidence on impact of the WFPP.	Michigan NRC monthly meeting.
May	Two Michigan DNR representatives visit Burlington, VT along with a University of Michigan soil scientist.	Letter writing, data collection, press releases to establish contacts and build membership for CRUF.	
June	Proponents lobby Michigan NRC for wood supply commitment.	CRUF representative attends wood fuel symposium at Central Michigan University. CRUF attends Michigan NRC meeting. "Demoralized" by prospects, directors of CRUF decide on waste issue as alternative strategy.	Michigan NRC monthly meeting.

Month			
July		CRUF representative attends Michigan NRC meeting in Ludington. Decision on allocation of wood from state forestland deferred. Local logger presents on side of opponents. Environmental lawyer is retained to draft waste ordinance for Hersey Township.	Michigan NRC monthly meeting.
August		CRUF representative attends Michigan NRC meeting. CRUF becomes a Michigan nonprofit corporation.	Michigan DNR memo recommends sale of wood from state land. Michigan NRC monthly meeting. Commissioners vow no decision will be made without a full environmental impact statement.
September		CRUF representative attends Michigan NRC public hearing on wood supply commitment.	Michigan NRC public hearing on wood supply commitment.
October		CRUF news conference reaffirming its stance against WFPP despite Michigan NRC decision on wood allocation. CRUF representatives interviewed on TV9/10 of Cadillac, "Eye on Michigan" show. Waste ordinance petition drive.	Michigan NRC decides to auction wood from state land instead of direct sale to utilities. Hersey Township Board meeting. Three hundred thirty signatures on petition presented by CRUF favoring waste ordinance. Ordinance is unanimously adopted.
November	Utilities threaten withdrawal, urge Hersey Township to reconsider ordinance, enter discussion of amending ordinance. Letters and press releases against waste ordinance.	CRUF representative attends Michigan Forest Association meeting on wood energy. CRUF lobbies to save waste ordinance. CRUF representative interviewed on WBRN radio. CRUF appeals to Osceola County Commission to oppose WFPP.	Burlington, VT WFPP shelved. Hersey Township board reaffirms its support of WFPP. Osceola County Commission proposes hearing on WFPP.

Appendix 1—*Continued*

Month	Promoters' Activities	Citizens' Activities	Relevant Hearings, Meetings, and Regulatory Activities
December	Consumers Power and Wolverine announce new Hersey project manager. Utility representative attends Osceola County Commisson meeting.	CRUF representatives and Hersey residents attend Osceola County Commission meeting. Request assurance for wood only in WFPP. CRUF is excluded from panel, mounts dissent from audience. CRUF continues fight to preserve waste ordinance.	Refuse-derived fuel (RDF) burning issue discussed at Osceola County Commission meeting. Commissioners urge legal document precluding use of RDF in the Hersey plant. Attorneys for utilities, CRUF, and Hersey Township meet in Traverse City to amend the toxic and hazardous waste ordinance.
1980			
January			Hersey Township board amends toxic and hazardous waste ordinance to require a permit for burning RDF.
March	Consumers Power presents contract for refuse-derived fuel test burn to Hersey Township board soon after amendment of ordinance.	CRUF representative attends meetings at Michigan State University's "Farmers Week." CRUF persuades Hersey Township board to hold special meeting on compliance contract issue.	Hersey Township board considers utilities' bid for RDF burn permit; postpones decision until April. Schedules a special meeting to vote on compliance contract.
April	Utilities announce decision to reevaluate Hersey site pursuant to Hersey Township vote on referendum. Proponents make presentation at special meeting of the Hersey Township board.	CRUF excluded from panel at special Hersey Township board meeting. Protests from audience. CRUF proposes referendum on contract issue at monthly meeting of Hersey Township board.	Special meeting of Hersey Township board to discuss compliance contract. Decision deferred. Local state representative grilled on WFPP at his town meeting in Evart. Hersey Township board votes unanimously at regular meeting for referendum on RDF burn permit.

May	Position statement by utilities published in *Osceola County Herald*. "Hersey Nixed as Wood Plant Site" is *Osceola County Herald* headline. Project Manager for Consumers Power attack CRUF's credibility in a series of letters in *Osceola County Herald*.	CRUF replies to each attack by WFPP promoters in *Osceola County Herald*.	Hersey Township board votes to withdraw referendum based on withdrawal of project by utilities. Resolves to reinstate referendum automatically should the utilities return.
June	Consumers Power announces receipt of unsolicited offers from ten communities desiring the WFPP. Consumers Power makes presentation in Evart at request of Chamber of commerce but declines invitation to present at Farwell meeting.	CRUF excluded from panel at Evart meeting; present at Farwell meeting on panel with Michigan DNR representative and University of Michigan soil scientist.	Evart meeting convened by Chamber of Commerce. Farwell meeting sponsored by local environmental groups.
August	Michigan DNR director defends the WFPP concept at a meeting in Bellaire.		Meeting in Bellaire to discuss Ellsworth site for WFPP.
September	Wolverine Electric withdraws from the venture, citing financing problems. Consumers Power announces a two to three year delay because of its financial situation and the withdrawal by Wolverine Electric.		
November		WFPP opponent Bion Jacobs unseats incumbent Hersey Township Supervisor Forest Benzing with active support from former CRUF members; two of four Hersey Township board members get elected with similar help. Former CRUF member narrowly defeated in race for Osceola County Commission seat despite carrying Hersey Township by a wide margin.	

Appendix 2:
Fact-Value Content Analysis Example

The following is an example of the way in which the fact/value typology discussed in chapter 7 was interpreted in the analysis of the Hersey wood-fired power plant debate. The account selected is a column that appeared in the *Osceola County Herald* on 1 November 1979 under the title "There's More than One Way to Fight Back." Numbers between slash marks indicate the end of an argument. The assigned fact-value scores are listed after the selection.

"If some individual was to cut down your trees, chip them, haul the chips into your back yard, mix them with trash which was also hauled into your back yard, then burn the whole mess for his very own profit, and then make you pay for the whole operation what would you do?

You would (I'll bet) balk like the proverbial "Missouri Mule." You would scratch and kick and bite and beller and stop the deal even if you had to use a shot gun.

But since it is a big corporation with plenty of money, and probably with some bureaucrats and politicians on their "subscription" list, and all of them soft soaping you with disguised half truths and half promises, you dear citizens are behaving like a trained puppy dog. You sit up, speak, roll over, or play dead on command. /1/

Many people believe the promise of 81 new jobs for the area.

If the power companies were sincere in that promise they would declare publicly that out of these 81 jobs at least 40 or 20 or maybe only 10 would be positively earmarked for the local people. They would give us a promise that "x number" of jobs would be filled by YOUR kids or grand kids.

But they will make no such promise. Three or four months ago Consumers, Morbark and Wolverine representatives made a presentation to the Osceola County Commissioners on the proposed plant. During the recess these men were asked about the number of jobs scheduled to be filled by local people. Their answers were evasive, with references to labor skills, technical requirements and special training. And when asked if even one, "MIND YOU, IF EVEN ONE LOUSY LITTLE JOB," could be guaranteed to some local boy or girl, each one of these three men suddenly remembered, that they had to catch some particular commissioner, and disappeared without answering the question.

They say that of these 81 jobs, 31 will be in-plant jobs. It is reasonable to expect that at least 20 of these jobs should be filled, not only available to, but "actually filled" by local young men and women. The short term expense of training these people would repay the plant owners a hundred-fold in local good will and agreeable acceptance. /2/

As a matter of fact instead of a job gain, there may be an eventual job loss in the county because of the use of improved heavy machinery. It was recently announced that because Morbark of Winn has outbid Buskirk Enterprises of Paris for the Menasha (Gaylord) chip contract, the Buskirk plant will close Dec. 1. This means a probable gain of two or three jobs for the city of Winn, but a loss of 10 jobs in Paris. /3/

The trash problem has been very well covered in previous articles in this paper, but up to the present time no responsible person has made a public statement about it. Now, Vernon J. Ehlers, chairperson of the Kent County Board of Commissioners (that's Grand Rapids) says that Consumers Power has "informally" agreed to burn some of this area's refuse (500 tons a day the feasibility study say; that's a lot of "some") in the woodburning power plant near Hersey. So now it's official. Pretty soon it will be an accomplished fact. /4/ And if we allow ourselves to be covered by the Grand Rapids and other big city trash we will certainly deserve it. /5/

On top of all of this they are going to add insult to injury by making you pay for the whole she-bang.

Yes sir! They are going to give you a rooking and make you pay for it too, while they, the investors, sit back and enjoy a tax-exempt income of maybe 10 or 12 or maybe even 18 percent on their investment. While you will not gain one penny, tax-wise. /6/ In fact, as Mr. Don Albosta stated in newspaper interview, YOU will pay for at least one new road and one new railroad siding for the plant.

But wait a dog-gone minute. Nobody has mentioned a new bridge for the new railroad siding. They have already rebuilt the highway bridges across the Muskegon River in Hersey and Paris, with your (State) money, but they will also need a bridge for the railroad to cross the river. And as per Mr. Albosta, WE will have to build it for them. /7/

It may be that all of this criticism is unwarranted, but before we allow this plant to be forced on us, we should demand that our elected officials tell us the whole truth. So let's all (this includes Mecosta, Osceola and Lake counties) call or write our city, township, county, state and federal elected officials and demand three things.

One: that at least 30 percent of all jobs generated by the plant be allocated to local people. /8/

Two: that no rubbish from outside the local area (Mecosta, Osceola, Lake) be burned either on a trial or permanent basis. /9/

Three: that no public money be used to assist, in any way, the building of the plant, or for highway or railroad construction to the plant. That only conventional financing be allowed. No lease leverage financing. /10/

POST SCRIPT. Just received a phone call from a CRUF member, that he has just received some newspaper clippings from a friend in Vermont. These clippings show that the Vermont plant, similar to the one proposed in Hersey, is to be shut down, because of high operating cost. It is cheaper to buy their power from Canada." /11/

Analysis of Arguments
(refer to fact-value scale in chapter 7)

1. Type 3: factual, normative apparent.
2. Type 3: factual, normative apparent.
3. Type 2: factual, normative implied.
4. Type 2: factual, normative implied.
5. Type 4: strictly normative.
6. Type 3: factual, normative apparent.
7. Type 3: factual, normative apparent.
8. Type 3: factual, normative apparent.
9. Type 3: factual, normative apparent.
10. Type 3: factual, normative apparent.
11. Type 3: factual, normative apparent.

The columnist was a nonexpert in the controversy and was coded accordingly.

Notes

Introduction

1. In addition to use of the original doctoral dissertation, six published articles on wood energy development and expertise were drawn together in order to write this book. Of the six, three received the benefit of presentation to an audience of researchers at national or international conferences. All six were transformed by peer and editorial review for publication in academic journals. As a result, the ideas presented here have been refined over a period of years by much discussion and a number of writings and revisions. The articles in question are: F. Frankena, "Facts, Values, and Technical Expertise in a Renewable Energy Siting Dispute," *Journal of Economic Psychology* 4 (1983): 131–47; F. Frankena, "Rethinking the Scale of Biomass Energy Conversion Facilities: The Case of Wood-Electric Power," *Biomass: An International Journal* 14, 4 (1987): 149–71; F. Frankena, "The Emergent Social Role and Political Impact of the Voluntary Technical Expert," *Environmental Impact Assessment Review* 8, 1 (1988): 73–84; F. Frankena, "Defeat of a Wood-Fired Electric Power Plant: A Study in Social Change," *Society and Natural Resources: An International Journal*, 1, 2 (1988): 167–83; F. Frankena, "Large-Scale Wood Energy Development in Michigan: An Assessment of Citizen Involvement," *Renewable Resources Journal*, 2, 1 (1989): 14–20; and F. Frankena, "Wood-fired Power Plants: Public Controversies Reveal Major Social Changes," *Journal of Forestry* 87, 4 (1989): 18–23.

2. J. M. Wilkes, "Case Studies; A Promising Way to Assess Technological Impacts? *4S Review* 1, 2 (Summer 1983): 12.

3. R. K. Yin, *Case Study Research: Design and Methods*, Vol. 5, Applied Social Research Methods Series (Beverly Hills, Calif.: Sage, 1984).

4. L. A. Coser, "Conflict: Social Aspects," *International Encyclopedia of the Social Sciences* (New York: Macmillan Co. and the Free Press, 1968), pp. 232–36; L. A. Coser, *The Functions of Social Conflict* (Glencoe, Ill.: The Free Press, 1956).

5. D. Nelkin and M. Pollak, "Ideology as Strategy: The Discourse of the Anti-Nuclear Movement in France and Germany," *Science, Technology and Human Values* 5 (Winter 1980): 3.

6. This need was pointed out by J. Ravetz, "Scientific Knowledge and Expert Advice in Debates about Large Technological Innovations," *Minerva* 16, 2 (Summer 1978): 273–82.

7. D. Nelkin, "The Political Impact of Technical Expertise," *Social Studies of Science* 5 (1975): 36–37. H. Nowotny, "Experts and Their Expertise: On the Changing Relationship between Experts and Their Public," *Bulletin of Science, Technology, and Society* 1 (1981): 235.

8. Ibid.; L. R. King and P. H. Melanson, "Knowledge and Politics: Some Experiences from the 1960's," *Public Policy* 20 (Winter 1972): 82–101; D. G. Jopling, S. J. Gage, and M. E. F. Schoeman, "Forecasting Public Resistance to

Technology: The Example of Nuclear Power Reactor Siting," in J. R. Bright and M. E. F. Schoeman (eds.), *A Guide to Practical Technological Forecasting* (Englewood Cliffs, N.J.: Prentice-Hall, 1973), pp. 53–66.

9. D. Nelkin, "Science and the Polity: Changing Relationships and Their Consequences for Social Studies of Science: Presidential Address—Fourth Annual Meeting," *Society for Social Studies of Science Newsletter* 4, 4 (Fall 1979): 6–7.

10. Michigan, State of, *Michigan Wood Energy Development Plan: An Addendum to Michigan's Forest Resources, A Statewide Resources Plan* (Lansing, Mich.: Department of Commerce and Department of Natural Resources, March 1986).

11. The difficulty of separating facts and values is discussed in J. DeSario and S. Langton, "Citizen Participation and Technocracy," in J. DeSario and S. Langton (eds.), *Citizen Participation in Public Decision Making* (New York: Greenwood Press, 1987), pp. 3–17. See also M. Enbar, "Equity in the Social Sciences," in R. E. Kasperson (ed.), *Equity Issues in Radioactive Waste Management* (Cambridge, Mass.: Oelgeschlager, Gunn and Hain, 1983), p. 23; A. S. Whittemore, "Facts and Values in Risk Analysis for Environmental Toxicants," *Risk Analysis* 3, 1 (March 1983): 23–33; D. Nelkin and M. Pollak, "Problems and Procedures in the Regulation of Technological Risk," in R. C. Schwing and W. A. Albers (eds.), *Societal Risk Assessment: How Safe Is Safe Enough?* (New York: Plenum Press, 1980), p. 247. R. Johnston ("Controlling Technology: An Issue for the Social Studies of Science," *Social Studies of Science* 14 (1984): 109) concludes that focusing on the separation of facts and values is a poor research strategy for studying the control of technology.

Chapter 1. The Politics of Expertise

1. T. Lowi in the preface of E. J. Feldman and J. Milch, *Technocracy Versus Democracy: The Comparative Politics of International Airports* (Boston, Mass.: Auburn House Pub. Co., 1982), p. xx.

2. M. Mulkay, *Science and the Sociology of Knowledge* (Boston: G. Allen and Uwin, 1979), p. 121.

3. B. Kennard, *Nothing Can Be Done, Everything Is Possible* (Andover, Mass.: Brick House Publishing Co., 1982), p. 139.

4. L. K. Caldwell, R. Hayes, and I. M. MacWhirter, *Citizens and the Environment: Case Studies in Popular Action* (Bloomington: Indiana University Press, 1976) provide an especially cogent discussion. They point out that the decision making process must be opened to the public simply because government agencies and energy conglomerates do not have all the experts. Much of the interest in public participation in technical decisions is related to energy technology development (e.g., Ibid.; J. W. Hendricks, "Public Participation and Democratic Decision Making on Energy Issues," *Social Society Energy Review*, 1, 2 (1978): 1–25; Ravetz, "Scientific Knowledge and Expert Advice in Debates about Large Technological Innovations").

5. Nelkin, "Science and the Polity"; K. G. Nichols, "The De-Institutionalization of Technical Expertise," in H. Skoie (ed.), *Scientific Expertise and the Public: Conference Proceedings* (Oslo: Institute for Studies in Research and Higher Education, Norwegian Research Council for Science and the

Humanities, 1979), pp. 35–48; Nelkin, "The Political Impact of Technical Expertise"; National Academy of Sciences, *Energy Choices for a Democratic Society*, Supporting Paper 7 (Washington, D. C.: Consumption, Location, and Occupational Patterns Resources Group Synthesis Panel, Committee on Nuclear and Alternative Energy Systems [CONAES], 1980), chapter 7; J. P. Holdren, "The Nuclear Controversy and the Limitations of Decision-Making by Experts," *Bulletin of the Atomic Scientists* 32 (March 1976): 20–22; I. R. Hoos, "The Credibility Issue," in *Essays on Issues Relevant to the Regulation of Radioactive Waste Management*, NUREG-0412 (Washington, D. C.: Office of Nuclear Material Safety and Safeguards, U. S. Nuclear Regulatory Commission, May 1978), pp. 21–30; G. Benveniste, *The Politics of Expertise*, 2nd ed. (Boston, Mass.: Boyd and Fraser Pub. Co., 1977); J. Primack and F. von Hippel, *Advice and Dissent: Scientists in the Political Arena* (New York: Basic Books, 1974); Jopling, Gage, and Schoeman, "Forecasting Public Resistance to Technology"; Nowotny, "Experts and Their Expertise," p. 239.

6. T. N. Gladwin, "Patterns of Environmental Conflict over Industrial Facilities in the United States, 1970–1980," *Natural Resources Journal* 20, 2 (1980): 243–74.

7. On this point S. P. Hays ("The Structure of Environmental Politics since World War II," *Journal of Social History* 14, 4 (Summer 1981): 719–38) observed: "This new source of political strength marked a change in the capacity of small-scale institutions to ward off intrusion from far above and beyond them." With regard to the success of counterexpertise in administrative politics, see S. P. Hays, *Beauty, Health, and Permanence: Environmental Politics in the United States, 1955–1985* (New York: Cambridge University Press, 1987), p. 315. R. H. K. Vietor (*Environmental Politics and the Coal Coalition* (College Station: Texas A & M University Press, 1980), pp. 233ff) also finds that after 1967 the complexities of environmental policy issues increasingly moved beyond the comprehension of nonexperts. See also Frankena, "The Emergent Social Role and Political Impact of the Voluntary Technical Expert."

8. Hays, *Beauty, Health, and Permanence*, p. 63.

9. Nichols, "The De-Institutionalization of Technical Expertise."

10. D. Nelkin, "Thoughts on the Proposed Science Court," *Newsletter on Science, Technology, and Human Values* No. 18 (January 1977): 20–31.

11. Nelkin, "The Political Impact of Technical Expertise," p. 36; Nowotny, "Experts and Their Expertise," p. 236; Hays, *Beauty Health, and Permanence*, p. 538. "Scientific knowledge, like land, labor, and capital, is a resource—indeed a commodity—and the ability to manipulate and control this resource has profound implications for the distribution of political power in democratic societies" (D. Nelkin, "Scientific Knowledge, Public Policy, and Democracy: A Review Essay." *Knowledge: Creation, Diffusion, Utilization* 1, 1 (September 1979): 118).

12. Nichols, "The De-Institutionalization of Technical Expertise," pp. 36–38; Hoos, "The Credibility Issue"; Mulkay, *Science and the Sociology of Knowledge*, p. 114: Nowotny, "Experts and Their Expertise," pp. 235–236, 238; A. G. Gross, "Public Debates as Failed Social Dramas: The Recombinant DNA Controversy," *Quarterly Journal of Speech* 70, 4 (1984): 407.

13. The following works support the generalizations made in this paragraph: Hays, *Beauty, Health, and Permanence*, pp. 206, 320; D. Nelkin, "Science, Technology, and Political Conflict: Analyzing the Issues," in D. Nelkin (ed.), *Controversy: The Politics of Technical Decisions*, 2nd ed. (Beverly Hills, Calif.: Sage, 1984), pp. 9–22; Hays, "The Structure of Environmental Politics since World War II," p. 724; T. Rozak, "Citadel of Expertise," chapter 3 in *Where the Wasteland Ends* (Garden City, N. Y.: Anchor Books, 1973), pp. 28–73; Frankena,

"The Emergent Social Role and Political Impact of the Voluntary Technical Expert"; Benveniste, *The Politics of Expertise*; Vietor, *Environmental Politics and the Coal Coalition*, p. 234; Feldman and Milch, *Technocracy Versus Democracy*, p. 146.

14. For example S. F. Tierney, "The Nuclear Waste Disposal Controversy," in D. Nelkin (ed.), *Controversy: The Politics of Technical Decisions*, 2nd ed. (Beverly Hills, Calif.: Sage, 1984), pp. 91–110; R. L. Crain, E. Katz, and D. B. Rosenthal, *The Politics of Community Conflict: The Fluoridation Decision* (New York: Bobbs-Merrill, 1969); Reich ("Environmental Politics and Science: The Case of PBB Contamination in Michigan," *American Journal of Public Health* 73, 3 (March 1983): 302–13) points out that in a polarized situation few scientists or scientific studies are perceived as "neutral."

15. Crain, Katz, and Rosenthal, *The Politics of Community Conflict*, pp. 68, 228; A. Mazur, *The Dynamics of Technical Controversy* (Washington, D. C.: Communications Press, 1981), p. 110; H. Henderson, *Creating Alternative Futures: The End of Economics* (New York: Berkley Publishing Corp., 1978), p. 294.

16. Ravetz, "Scientific Knowledge and Expert Advice in Debates about Large Technological Innovations," pp. 280–281.

17. Ibid.

18. I. D. Clark, "Expert Advice in the Controversy about Supersonic Transport in the United States," *Minerva* 12 (October 1974): 416.

19. F. Frankena, *Experts and Expertise in Environmental Litigation: A Bibliography*, P-1909 (Monticello, Ill.: Vance Bibliographies, April 1986).

20. Hays, "The Structure of Environmental Politics since World War II"; Gladwin, "Patterns of Environmental Conflict over Industrial Facilities in the United States, 1970–1980." The term "local expert" is used by B. C. Aldrich, *Communities of Opposition: Energy Facility Siting in Minnesota* (Winona, Minn.: Winona State University, 1980), p. 58.

21. With regard to the ideology and biases of particular disciplines and specialities, see: I. G. Barbour, *Technology, Environment, and Human Values* (New York: Praeger, 1980), p. 201; Feldman and Milch, *Technocracy Versus Democracy*, p. 139; R. V. Bartlett, *The Reserve Mining Controversy: Science, Technology, and Environmental Quality* (Bloomington, Ind.: Indiana University Press, 1980), p. 214; Hays, *Beauty, Health, and Permanence*, p. 357; J. Rayner and D. Peerla, "To Spray or Not to Spray? Exclusive Concepts of Science and Nature in the Canadian Spruce Budworm Controversy," paper presented at the First National Symposium on Social Science in Resource Management, Corvallis, Oreg., May 1986.

22. Nelkin, "Science, Technology, and Political Conflict."

23. Feldman and Milch, *Technocracy Versus Democracy*, p. 175.

24. Ibid., p. 131.

25. Hays, "The Structure of Environmental Politics since World War II," p. 728.

26. Aldrich, *Communities of Opposition*, p. i.

27. The changing scale of social activities and their environmental impact is discussed in B. Stokes, *Helping Ourselves: Local Solutions to Global Problems* (Washington, DC: Worldwatch Institute, 1981), and in the following three works by S. P. Hays: "The Structure of Environmental Politics since World War II"; "Value Premises for Planning and Public Policy: The Historical Context," in R. N. L. Andrews (ed.), *Land in America: Commodity or Natural Resource?* (Lexington, Mass.: Lexington Books, 1979), pp. 149–66; *Beauty, Health, and*

Permanence. The quote at the end of the paragraph is from Nelkin, "Science and the Polity," pp. 6–7.

28. If the general study of expertise is in need of development (Mulkay, *Science and the Sociology of Knowledge*, p. 121), then the volunteer expert is manifestly an unexplored subject. The diversity of descriptive terms—"counter-expertise," "citizen experts," "local experts," "countervailing expertise," and "volunteer activists" (see Frankena, "The Emergent Social Role and Political Impact of the Voluntary Technical Expert" for the sources which adopted these terms)—suggest a lack of consensus about the subject. R. Mitchell ("From Elite Quarrel to Mass Movement," *Society* 18, 5 (1981): 76–84) and Mazur (*The Dynamics of Technical Controversy*) make use of the term "volunteer expert," the one preferred here. A check of the Permuterm Index of the *Social Sciences Citation Index* for the years 1966–1984 yielded not one citation that meaningfully conjoined "volunteer" or "voluntary" with "expert" or its variants.

29. Nelkin, "The Political Impact of Technical Expertise." Other researchers who have noted the success of groups having unequal expertise include Gladwin, "Patterns of Environmental Conflict over Industrial Facilities in the United States, 1970–1980"; F. Frankena, *The Impact of Technical Expertise in a Nonmetropolitan Siting Dispute: A Case Study of the Hersey Wood-Fired Power Plant Controversy*, Ph. D. dissertation, Michigan State University, East Lansing, Mich., 1982; Benveniste, *The Politics of Expertise*; Primack and von Hippel, *Advice and Dissent*; Jopling, Gage, and Schoeman, "Forecasting Public Resistance to Technology"; Aldrich, *Communities of Opposition*; Hays, "The Structure of Environmental Politics since World War II," p. 724; Hays, *Beauty, Health, and Permanence*, p. 315.

30. D. Goodman, "Ecological Expertise," in H. A. Feiveson et al. (eds.), *Boundaries of Analysis: An Inquiry Into the Tocks Island Dam Controversy* (Cambridge, Mass.: Ballinger, 1976), pp. 354–55; Barbour, *Technology, Environment, and Human Values*, p. 170; Hays, *Beauty, Health, and Permanence*, p. 530.

H. Nowotny ("The Role of the Experts in Developing Public Policy: The Austrian Debate on Nuclear Power," *Science, Technology, and Human Values* 5 (Summer 1980): 17) describes this interaction of facts and values as "conflict by proxy." Her description of the behavior of experts in this light is well worth quoting: "The experts argued *as if* the underlying issues could be resolved by rational debate alone. No one was prepared to draw the more obvious conclusion that the conflict could not be solved by experts, since they were merely reproducing arguments that had already been presented in the normative-political arena. Bound by their self conceptions and their public roles as experts, they were forced to act *as if* they were engaged in scientific argument, while they were instead carrying out a conflict by proxy."

31. That is why the science court cannot be expected to resolve controversies (Nelkin, "Thoughts on the Proposed Science Court"; seconded by Ravetz, "Scientific Knowledge and Expert Advice," p. 281). On the inseparability of facts and values, see note 11 in the Introduction. The social utility of controversies as a means of conflict resolution and social learning has been promoted by A. Rip ("Controversies as Informal Technology Assessment," *Knowledge: Creation Diffusion, Utilization* 8, 2 (1986): 347–71), an idea that makes good sense by this reckoning of values and facts.

32. Both Kennard and Barbour discuss the notion that we are all experts when it comes to values. Kennard, *Nothing Can Be Done, Everything Is Possible*;

Barbour, *Technology, Environment, and Human Values.* See quotation from Klessig and Strite at the beginning of chapter 10, p. 215.

33. Nelkin, "Thoughts on the Proposed Science Court," p. 22.

34. Nichols, "The De-Institutionalization of Technical Expertise" p. 38; D. Nelkin, "The Role of Experts in a Nuclear Siting Controversy," *Bulletin of the Atomic Scientists* 30 (November 1974), p. 35; Frankena, *The Impact of Technical Expertise in a Nonmetropolitan Siting Dispute*; Chapter 5; Frankena, "Facts, Values, and Technical Expertise in a Renewable Energy Siting Dispute."

35. Nelkin, "The Political Impact of Technical Expertise," pp. 53–54, W. Edwards and D. von Winterfeldt, "Public Disputes about Risky Technologies: Stakeholders and Arenas," in V. Covello, J. Menkes, and J. Mumpower (eds.), *Risk Evaluation and Management* (New York: Plenum Press, 1984), pp. 86–87.

36. Hays, *Beauty, Health, and Permanence*, p. 478.

37. Feldman and Milch, *Technocracy Versus Democracy*, p. 131.

38. Hays, "Value Premises for Planning and Public Policy," p. 150.

39. Reich, "Environmental Politics and Science," p. 312.

40. Feldman and Milch, *Technocracy Versus Democracy*, p. 24.

41. R. Sclove, "Decision-Making in a Democracy," *Bulletin of the Atomic Scientists* (May 1982): 46.

42. Nowotny, "Experts and Their Expertise," p. 239.

43. Feldman and Milch, *Technocracy Versus Democracy*, p. 141.

44. Ibid., p. 227.

45. Benveniste, *The Politics of Expertise.*

46. Nelkin, "Science, Technology, and Political Conflict," p. 20.

Chapter 2. Social Change as a Source for Controversy in Wood Energy Development

1. A. Sokolow, "Local Politics and the Turnaround Migration: Newcomer-Oldtimer Relations in Small Communities," in C. Roseman, A. Sofranko, and J. Williams (eds.), *Population Redistribution in the Midwest* (Ames, Iowa: North Central Regional Center for Rural Development, Iowa State University, 1981), pp. 169–90; C. Cortese, "Rapid Growth and Social Change in Western Communities," *Social Impact Assessment* 40/41 (April–May 1979): 1–7; *Institute for Social Research Newsletter*, "Local Officials Must Plan for Growth, Northern Michigan Residents Declare," 6, 3 (1978): 2, 8.

With regard to the tendency for cooperation rather than conflict between old timers and newcomers in turnaround communities, see Sokolow ("Local Politics and the Turnaround Migration") and D. J. Blahna "Social Bases for Resource Conflicts in Areas of Reverse Migration" in R. G. Lee, D. R. Field, and W. R. Burch, Jr., (eds.), *Community and Forestry: Continuities in the Sociology of Natural Resources* (Boulder, CO: Westview Press), pp. 159–178. The latter study is especially relevant because it was conducted in northern lower Michigan, the region containing Hersey and Indian River. Blahna also finds that the turnaround is responsible for bringing political sophistication to nonmetropolitan communities.

2. Hays, "The Structure of Environmental Politics since World War II," p. 720.

3. S. L. Albrecht, "Socio-Cultural Factors and Energy Resource Development

in Rural Areas of the West," *Journal of Environmental Management* 7, 1 (1978): 73–90.

4. An especially cogent statement on the role of the environmental movement in U.S. politics is made by Hays in *Beauty, Health, and Permanence.* A major thesis of this history is the centrality of value change as a basis for the emergence of U.S. environmental politics (2ff, 527ff).

5. PBB is the acronym for polybrominated biphenyl, a complex organic chemical used as a fire retardant. It was first produced in the 1970s, as noted in J. Egginton, *The Poisoning of Michigan* (New York: W. W. Norton and Co., 1980). See Reich ("Environmental Politics and Science," 309ff) for a study of the interaction between science and politics in the Michigan PBB contamination.

6. See B. W. Coyer and D. S. Schwerin ("Bureaucratic Regulation and Farmer Protest in the Michigan PBB Contamination Case," *Rural Sociology* 46, 4 (Winter 1981): 703–23) for a political case study of dissatisfaction among farmers and the response by regulatory agencies.

7. E. Chen, *PBB: An American Tragedy* (Englewood Cliffs, N. J.: Prentice-Hall, 1979), and Egginton, *The Poisoning of Michigan.*

8. See Coyer and Schwerin, "Bureaucratic Regulation and Farmer Protest in the Michigan PBB Contamination Case."

9. Egginton, *The Poisoning of Michigan.*

10. Chen, *PBB, An American Tragedy.*

11. Egginton, *The Poisoning of Michigan.*

12. Ibid.

13. Ibid.

14. *Osceola County Herald*, "RDF Hot Issue in Hersey," April 17, 1980, 1, 4B.

15. B. M. Casper and P. D. Wellstone, *Powerline: The First Battle of America's Energy War* (Amherst, Mass.: University of Massachusetts Press, 1981), p. 302.

16. Examples of siting disputes for the facilities mentioned include J. Milch, "The Toronto Airport Controversy," in D. Nelkin (ed.), *Controversy: The Politics of Technical Decisions*, 2nd ed. (Beverly Hills, Calif.: Sage, 1984), pp. 27–50, and Feldman and Milch, *Technocracy Versus Democracy* [airports]; Casper and Wellstone, *Powerline* [power transmission lines]; Tierney, "The Nuclear Waste Disposal Controversy" [radioactive waste disposal facilities]; Aldrich, *Communities of Opposition* [coal-fired power plants]; Jopling, Gage, and Schoeman, "Forecasting Public Resistance to Technology"; D. Nelkin, *Nuclear Power and Its Critics: The Cayuga Lake Controversy*, Science, Technology and Society Series No. 1 (Ithaca, N. Y.: Cornell University Press, 1971) which is summarized and updated in D. Nelkin, "Nuclear Power and Its Critics: A Siting Dispute" in D. Nelkin (ed.), *Controversy: The Politics of Technical Decisions*, 2nd ed. (Beverly Hills, Calif.: Sage, 1984), pp. 51–71 [nuclear power plants]; and, L. L. Klessig and V. L. Strite, *The ELF Odyssey: National Security Versus Environmental Protection* (Boulder, Colo.: Westview Press, 1980) and S. L. Albrecht, "Community Response to Large-Scale Federal Projects: The Case of the MX," in S. H. Murdock et al. (eds.), *Nuclear Waste: Socioeconomic Dimensions of Long-term Storage* (Boulder, Colo.: Westview Press, 1983), pp. 233–50 [defense facilities].

17. An extensive review of the siting literature is beyond the scope of this study. C. Cluett, M. Greene, and L. Radford, *Individual and Community Response to Energy Facility Siting: A Review of the Literature.* B–HARC–411–045 (Seattle, Wash.: Battelle Affairs Research Centers, November 1979) provide a thorough but dated review on community response to energy facility siting in all phases of the

development process. An annotated bibliography is appended to the review, reproduced with minor abridgement in C. Cluett, M. Greene, and L. Radford, *Individual and Community Response to Energy Facility Siting: An Annotated Bibliography*, Public Administration Series no. 493 (Monticello, Ill.: Vance Bibliographies, May 1980).

18. C. P. Wolf, "The NIMBY Syndrome: Its Cause and Cure," in F. Sterrett (ed.), *Environmental Sciences* (New York: New York Academy of Sciences, 1987), pp. 216–29.

19. A. Jakimo and I. C. Bupp, "Nuclear Waste Disposal: Not in My Backyard," *Technology Review* 80, 5 (1978): 64–72; E. J. Farkas ("The NIMBY Syndrome," *Alternatives* 19, 2–3 (1982): 47–50) apparently was the first to qualify it as a syndrome.

20. M. Harthill (ed.), *Hazardous Waste Management: In Whose Backyard?* AAAS Selected Symposium Series (Boulder, Colo.: Westview Press, 1984); Institute for Environmental Negotiation (ed.), *Not-in-My-Backyard!: Community Reaction to Locally Unwanted Land Use* (Charlottesville, Va., 1985); A. R. Matheny and B. A. Williams, "Knowledge vs. NIMBY: Assessing Florida's Strategy for Siting Hazardous Waste Disposal Facilities, "*Policy Studies Journal* 14, 1 (1985): 70–80; Wolf, "The NIMBY Syndrome."

21. Wolf, "The NIMBY Syndrome," p. 217.

22. Ibid.

23. Hays, *Beauty, Health, and Permanence*, p. 142.

24. The classic studies of the turnaround include C. L. Beale, "The Recent Shift of U.S. Population to Nonmetropolitan Areas, 1970–75," *International Regional Science Review* 2, 2 (1975): 113–22, and *The Revival of Population Growth in Nonmetropolitan America* (Washington, D.C.: Economic Research Service, USDA, 1975); P. A. Morrison and J. P. Wheeler, "Rural Renaissance in America? The Revival of Population Growth in Remote Areas," *Population Bulletin* 31, 3 (1976), entire issue. With regard to the effect of the population migration turnaround in the Great Lakes Region, see G. V. Fuguitt and C. L. Beale, "Post-1970 Shifts in the Pattern of Population Change in the North Central Region," in J. A. Beegle and R. L. McNamara (eds.), *Patterns of Migration and Population Change in America's Heartland* (East Lansing, Mich.: Michigan Agricultural Experiment Station, 1978), pp. 14–21, and P. R. voss and G. V. Fuguitt, *Profile: The Region's New Residents* (Washington, D. C.: Upper Great Lakes Regional Commission, U.S. Department of Commerce, December 1979).

25. J. Herbers, *The New Heartland* (New York: Times Books, 1986), p. 13.

26. The survey cited is Voss and Fugitt, *Profile: The Region's New Residents*. Motives for the return migration to rural areas are discussed in: L. A. Ploch, "The Reversal in Migration Patterns—Some Rural Development Consequences," *Rural Sociology* 43, 2 (Summer 1978): 293–303; K. F. McCarthy and P. A. Morrison, "The Changing Demographic and Economic Structure of Nonmetropolitan Areas of the United States," *International Regional Science Review* 2, 2 (1977): 123–42; F. Frankena, *Community Impacts of Rapid Growth in Nonmetropolitan Areas: A Literature Survey*, Rural Sociology Series no. 9. (East Lansing, Mich.: Michigan Agricultural Experimental Station, June 1980); Voss and Fuguitt, *Profile: The Region's New Residents*; R. D. Kahoe, "Motivations for Urban-Rural Migration," *Journal of Social Psychology* 96, 2 (1975): 303–4; L. H. Long and K. A. Hansen, *Reasons for Interstate Migration: Jobs, Retirement, Climate, and Other Influences*, Current Population Reports, Special Studies Series P-23, no. 81 (Washington, D.C.: U.S. Bureau of Census, March 1979); R. W. Marans and J. D. Wellman, *The*

Quality of Nonmetropolitan Living: Evaluations, Behaviors, and Expectations of Northern Michigan Residents (Ann Arbor, Mich.: Survey Research Center, 1978); J. D. Williams and A. J. Sofranko, " Motivations for the Immigration Component of Population Turnaround in Nonmetropolitan Areas," *Demography* 16, 2 (1979): 239–55.

27. Voss and Fuguitt, *Profile: The Region's New Residents*.

28. American Society of Planning Officials, *Subdividing Rural America: Impacts of Recreational Lot and Second Home Development*, prepared for the Council on Environmental Quality (Washington, D. C.: Government Printing Office, 1976).

29. Hays, "Value Premises for Planning and Public Policy," p. 162; Hays, *Beauty, Health, and Permanence*, p. 431.

30. The classic statement on this subject is Ploch, "The Reveral in Migration Patterns," A study of the county in which the Hersey controversy took place shows in precise terms the connection between amenities and turnaround-related growth. See F. Frankena and T. Koebernick, "The Pattern of Recent Housing Growth in a Nonmetropolitan County: Effects of Environment and Location," *Growth and Change* 15, 4 (1984): 32–42.

31. Aldrich, *Communities of Opposition*.

32. Hays, *Beauty, Health, and Permanence*, p. 432.

33. Cortese, "Rapid Growth and Social Change in Western Communities."

34. Albrecht, "Community Response to Large-Scale Federal Projects," p. 242. Albrecht's observation accords with that of D. Nelkin ("Controversy as a Political Challenge," in B. Barnes and D. Edge (eds.), *Science in Context: Readings in the Sociology of Science* (Cambridge, Mass.: MIT Press, 1982), p. 276), who finds that most active members of social groups in science and technology controversies "are middle-class, educated people with sufficient economic security and political skill to participate in decision making."

35. J. A. Christenson, "Value Configurations for Ruralites and Urbanites: A Comment on Bealer's Paper," *Rural Sociologist* 1, 1 (January 1981): 42–47.

36. K. Richter (*Nonmetropolitan Growth in the Late 1970s: The End of the Turnaround?* Working Paper 83–20 (Madison, WI: Center for Demography and Ecology, University of Wisconsin, 1983)) finds a similar downturn for the United States. However, it would be premature to project an end to the turnaround. Despite energy crises and recessions the turnaround continues with vigor in many counties. The nonmetropolitan turnaround may well turn out to be for the demographers and rural sociologists of the present what the post-war baby boom was to their depression-era counterparts. For a discussion of long-term ecological factors that might serve to continue and accelerate the turnaround, see F. Frankena, "Regional Socioeconomic Impacts of Declining Net Energy," *Urban Ecology* 3, 2 (1978): 101–10.

37. D. J. Blahna, "Rural Population Growth and Social Information Needs of Forest Management Agencies," in L. N. Wenner (ed.), *Social Science Information and Resource Management: Proceedings from an Interagency Symposium* (Washington, D. C.: U. S. Forest Service, 1985), pp. 23–31; Herbers, *The New Heartland*.

38. Herbers, *The New Heartland*, pp. 9–10.

39. Albrecht, "Community Response to Large-Scale Federal Projects."

40. The quotation is from Hays, "Value Premises for Planning and Public Policy," p. 161.

41. Works used in support of this view of the conflicting ideologies of development and autonomy include: Cluett, Greene, and Radford, *Individual and*

Community Response to Energy Facility Siting, 1979; Cortese, "Rapid Growth and Social Change in Western Communities"; Nelkin, "Science, Technology, and Political Conflict"; P. J. Tichenor, G. A. Donohue, and C. N. Olien, *Community Conflict and the Press* (Beverly Hills, Calif.: Sage Publications, 1980); Aldrich, *Communities of Opposition*; Hays, *Beauty, Health, and Permanence*; J. W. Gartrell, "Community as a Social Collective," in G. F. Summers and A. Selvik (eds.), *Energy Resource Communities* (Madison, Wis.: MJM Publishing Co., 1982), pp. 199–218.

Casper and Wellstone (*Powerline,* 302ff) regard loss of control to be the overriding issue for communities threatened by development of large-scale facilities.

42. Ideas were taken from the following works to characterize the developmental ideology of project promoters: H. Molotch, "Oil in Santa Barbara and Power in America," *Sociological Inquiry* 40 (Winter 1970): 131–44; National Academy of Sciences, *Energy Choices for a Democratic Society*; Nelkin, "Nuclear Power and Its Critics"; Aldrich, *Communities of Opposition*; Hoos, "The Credibility Issue."

43. See Mazur ("Opposition to Technological Innovation, "*Minerva* 13 (Spring 1975): 58–81) for a discussion of the connection between local disputes and the rise and fall of interest in values of broad national significance. Particularly relevant to this case study is the anomaly he finds between rise of the environmental movement and opposition to nuclear power plants but not to fossil fuel power plants. Mazur cites evidence that fossil fuel plants pollute more. He suggests that novelty is the factor causing greater public concern about nuclear power plants. It may be that novelty is also a factor in the public response to large-scale wood energy development.

The need for research on different kinds of power plants is suggested by Cluett, Greene, and Radford, *Individual and Community Response to Energy Facility Siting,* and R. S. Krannich, "Socioeconomic Impacts of Power Plant Development on Nonmetropolitan Communities: An Analysis of Perceptions and Hypothesized Impact Determinants in the Eastern United States," *Rural Sociology* 46 (Spring 1980): 128–42.

44. These ideas are proposed, respectively, by I. L. Horowitz ("Sociological and Ideological Conceptions of Industrial Development," *American Journal of Economics and Sociology* 23 (October 1964): 351–74) and Nelkin ("Science, Technology, and Political Conflict").

45. State environmental planning is often in conflict with rural areas, reinforcing a local perspective on environmental quality and encouraging the use of local autonomy as an issue. Energy companies have routinely exploited the power of state governments to enforce siting via state programs that override local veto power. At the same time the political context of siting is "the city reaching out to the larger region to use it for its own purposes." Given these two power tendencies, the rise of local autonomy is not surprising. See Hays, "Value Premises for Planning and Public Policy," pp. 159–163. The term "industrial invasion" has also been applied to the threat posed by siting (Aldrich, *Communities of Opposition*, p. 10).

46. Aldrich, *Communities of Opposition*, quoted on p. 24.

47. Regarding the cosmopolitan perspective of local leaders, see Mazur, "Opposition to Technological Innovation"; Aldrich, *Communities of Opposition*; and T. A. Caine, "The Willmar Bank Strike," in R. P. Wolensky and E. J. Miller (eds.), *The Small City and Regional Community: Proceedings of the Conference on the Small City and Regional Community*, vol. 2 (Stevens Point, Wis.: Foundation Press, Inc., 1979), pp. 158–66. Arguments in favor of WFPPs as a decentralized technology are advanced by M. Harris ("Vermont: Goodbye Coal, Hello Trees,"

Mother Jones 3 (December 1978): 13–4) and, in connection with the Hersey controversy, by W. H. Sells, "Use of Wood as and Energy Source," *in Proceedings of the 1979 Energy Information Forum and Workshop for Educators* (Lansing, Mich.: Michigan Educators Energy Forum, 1979), p. 91–95.

48. Caldwell, Hayes, and MacWhirter, *Citizens and the Environment*, pp. xiii-xiv.

Chapter 3. Wood Energy Development: Concepts and Issues

1. See C. C. Burwell, "Solar Biomass Energy: An Overview of U. S. Potential," *Science* 199 (10 March, 1978): 1041–48.

2. N. Georgescu-Roegen, *The Entropy Law and the Economic Process* (Cambridge, Mass.: Harvard University Press, 1971), p. 232.

3. W. R. Catton, Jr. "Depending on Ghosts," *Humboldt Journal of Social Relations* 2 (Fall/Winter 1974): 45–49.

4. R. Deis, "Where There's Wood, There's Smoke: Are We Being Burned by Burning Wood?" *Environmental Action* 12 (December 1980): 5.

5. C. E. Hewett and E. Peterson, "The Forest Resource: Emerging Conflicts and the Need for Action," in C. E. Hewett and T. E. Hamilton (eds.), *Forests in Demand: Conflicts and Solutions* (Boston, Mass.: Auburn House Pub., 1982), p. 17ff.

6. A. Grace, *Biomass: Solar Energy from Farms and Forests*, prepared for the Solar Energy Research Institute (Washington, D. C.: U.S. Government Printing Office, 1980).

7. J. Veigel and J. H. Lohnes, "The Forest and Energy," in C. E. Hewett and T. E. Hamilton (eds.), *Forests in Demand: Conflicts and Solutions* (Boston, Mass.: Auburn House Pub., 1982), p. 45.

8. C. E. Hewett et al., "Wood Energy in the United States," *Annual Review of Energy* 6 (1981): 142.

9. Deis, "Where There's Wood There's Smoke," p. 5.

10. Hewett and Peterson, "The Forest Resource," p. 14; Hewett et al., "Wood Energy in the United States," p. 140.

11. *Michigan Out-of-Doors*, "Wood To Provide Energy for Small Town," (May 1978): 113.

12. The Burlington, Vermont love affair with wood-electric power is detailed in chapter 5. It is the classic case that serves to illustrate the perils and pitfalls of large-scale wood energy development.

13. Hewett and Peterson ("The Forest Resource," p. 14) note that the value of the forest is changing, moving beyond forests as a supplier of materials to include recreation, wildlife habitat, watershed integrity, and the like.

14. The U.S. Department of Energy opted for large-scale wind generators in 1978. A demonstration project in Rhode Island resulted in technical problems and persistent community opposition. Such opposition has also arisen in California's Altamont Pass because of the constant noise of the wind generators dotting the landscape. See K. Black, "Tilting at Windmills," *Northwest Magazine* 18, 1 (1987): 13–16.

15. Casper and Wellstone (*Powerline*, p. 307ff) argue that schemes like the solar satellite system are the epitome of large-scale approaches to solar development requiring sacrifices for rural people.

16. Hewett et al., "Wood Energy in the United States," p. 142.

17. Ibid., pp. 152, 146.

18. D. Pimental, S. Chick, and W. Vergara, "Energy from Forests: Environmental and Wildlife Implications," *Interciencia* 6 (1981): 329–35.

19. U.S. Environmental Protection Agency, *Preliminary Environmental Assessment of Biomass Conversion to Synthetic Fuels*, EPA–600/778–204 (Cincinnati, Ohio: Battelle Columbus Laboratory for EPA/IERI, October 1978).

20. J. A. Cooper, "Environmental Impact of Residential Wood Combustion Emissions and Its Impact," *Journal of the Air Pollution Control Association* 30 (1980): 855–61.

21. U.S. Department of Energy, *Environmental Readiness Document—Wood Combustion*, DOE/ERD—0026 (Washington, D.C.: Assistant Secretary for the Environment, August 1979).

22. Deis, "Where There's Wood There's Smoke," p. 6.

23. Hewett et al., "Wood Energy in the United States," p. 160.

24. This listing is an adaptation of work by T. P. Huber, E. C. Gruntfest, and L. Lapalme-Roy, "The Use of Wood as Fuel in North America: Prospects and Problems," *Journal of Environmental Systems* 14, 4 (1984–85): 321–32.

25. Hewett and Peterson, "The Forest Resource," p. 15.

26. In an unusual study, A. Shama and K. Jacobs (*Social Values and Solar Energy Policy: The Policymaker and the Advocate* (Golden, Colo.: Solar Energy Research Institute, 1980(?)) compare and contrast the values of policymakers and advocates in connection with solar energy development. They found that policymakers emphasize the economic and national security values of solar energy, while solar energy advocates stress environmental, ethical, and social values. These contrasting values are a clear parallel with this study. The conditions for controversy are in place when decision makers do not assimilate the values of affected communities.

27. A. B. Lovins, "Energy Strategy: The Road Not Taken." *Foreign Affairs* 55, 1 (1976): 65–96; see also A. B. Lovins, "Soft Energy Technologies," *Annual Review of Energy* 3 (1978): 477–517. A further measure of the importance of the debate is the outpouring of research and writing on its various facets. See, for example, M. Messing, M. O'Meara, and R. M. Hall, *Report on the Jurisdictional Authorities of State and Local Government Related to Centralized and Decentralized Alternative Energy Systems* (Washington, D.C.: Environmental Policy Institute, November 1976), and the bibliography by J. M. Ohi et al., *Decentralized Energy Studies*, SERI/RR–774–448 (Golden, Colo.: Solar Energy Research Institute, May 1980).

28. U.S. Select Committee on Small Business and the Committee on Interior and Insular Affairs, *Alternative Long Range Energy Strategies: Hearings* (Washington, D.C.: U.S. Government Printing Office, 1976). Appendices in a separate volume under the same title total 1,553 pages (U.S. Select Committee on Small Business and the Committee on interior and Insular Affairs, *Alternative Long Range Energy Strategies: Appendices* (Washington, D.C.: U.S. Government Printing Office, 1976).

29. Casper and Wellstone, *Powerline*, p. 306.

Chapter 4. A History of the Hersey Controversy

1. M. Harris, "The Boom in Wood Use: Promise or Peril?" *American Forests* 86 (1980): 34–36.

2. Michigan, State of, *Michigan Wood Energy Development Plan*, p. 11.

3. M. Hiser et al., *Wood Fueled Power Generation: A Potential Source of Energy for Northern Michigan* (Lansing, Mich.: Michigan Public Service Commission, November 1977).

4. M. L. Hiser (ed.), *Wood Energy: Proceedings of Governor William G. Milliken's Conference, November 29, 1977, Ann Arbor, Michigan* (Ann Arbor, Mich.: Ann Arbor Science Publishers, 1978).

5. W. G. Milliken, "Keynote Address," in M. L. Hiser (ed.), *Wood Energy: Proceedings of Governor William G. Milliken's Conference, November 29, 1977* (Ann Arbor, Mich.: Ann Arbor Science Publishers, 1978), pp. 11–12.

6. *Evart Review*, "Reagan Lauds Morbark," 4 January 1979, p. 14.

7. J. Hale, "Electric Generating Plant under Study," *Big Rapids Pioneer*, 6 June 1978, p. 1A.

8. Daverman and Associates, *A Feasibility Study for a Waste-Wood Electric Generating Plant*.

9. For further details of this network, see Frankena, *The Impact of Technical Expertise in a Nonmetropolitan Siting Dispute*.

10. *Lansing State Journal*, "1983 Target for Wood-Fired Plant," 6 February 1979, p. 4B.

11. Sells, "Use of Wood as an Energy Source," p. 94.

12. Ibid.

13. Daverman and Associates, *A Feasibility Study for a Waste-Wood Electric Generating Plant*, prepared for Consumers Power Company, Morbark Industries, Inc., and Wolverine Electric Cooperative (Grand Rapids, Mich., February 1979), Appendix F.

14. *Grand Rapids Press*, "Hersey Does Slow Burn over Proposed Wood-Burning Plant," 21 January 1979, p. 1E.

15. Daverman and Associates, *A Feasibility Study for a Waste-Wood Electric Generating Plant*, Executive Summary.

16. Daverman and Associates, *A Feasibility Study for a Waste-Wood Electric Generating Plant*.

17. Ibid.

18. Ibid.

19. Ibid., Executive Summary.

20. See quote in chapter 8, p. 201, describing how the citizen group opposing the Hersey WFPP (Committee for the Rational Use of Our Forests) acquired the expertise to contest the project.

21. Daverman and Associates, *A Feasibility Study for a Waste-Wood Electric Generating Plant*, Appendix I, pp. 106–107.

22. Ibid., Executive Summary.

23. Daverman and Associates, *A Feasibility Study for a Waste-Wood Electric Generating Plant*.

24. *Osceola County Herald*, "Hersey Site Chosen for Chipping Plant," 28 September 1978, p. 1A.

25. *Grand Rapids Press*, "Osceola Will Get Wood-Burning Power Plant," 13 January 1979, p. 8B.

26. For a review of relevant literature and a study of locational determinants of growth in Osceola County, see Frankena and Koebernick, "The Pattern of Recent Housing Growth in a Nonmetropolitan County."

27. Ibid.

28. R. W. Rathge, *The Institutional Impacts of Rapid Population Growth on a Nonmetropolitan Michigan County*, Ph. D. dissertation, Michigan State University, East Lansing, Mich., 1981.

29. Tichenor, Donohue, and Olien, *Community Conflict and the Press*, pp. 94–95.

30. Aldrich, *Communities of Opposition*.

31. Those seeking a more quantitative analysis of the turnaround should consult Frankena and Koebernick, "The Pattern of Recent Housing Growth in a Nonmetropolitan County" and Blahna, "Rural Population Growth and Social Information Needs of Forest Management Agencies"; "Social Bases for Resource Conflicts in Areas of Reverse Migration." The latter two works provide survey evidence on the turnaround as a social basis for conflict in northern lower Michigan.

32. *Osceola County Herald*, "Opinions Split on Plant Near Hersey," 26 October 1978, pp. 1, 3A.

33. *Osceola County Herald*, "Hersey Residents Plan Meeting," 14 December 1978, p. 1A.

34. *Osceola County Herald*, "Hersey Residents Not Sold on Plant," 21 December 1978, p. 1A.

35. Ibid.

36. Ibid.

37. *Osceola County Herald*, "Young Explains Success of Vermont Plant," 25 January 1979, p. 7A. "Wood Chip Executive Speaks Out: Hersey Still Probable Site," 31 January 1979, pp. 3, 5.

38. *Detroit Free Press*, "Town Debates Wood-Power Plant," 13 February 1979, pp. 3, 11A.

39. D. Bolyard, "Letters to the Editor: Hersey Plant," *Michigan Out-of-Doors* 35 (May 1979): 6, 8.

40. *Osceola County Herald*, "200 Attend Hersey Wood Chip Plant Meeting," 15 February 1979, p. 1A.

41. Ibid.; *Osceola County Herald*, "Hersey Woodchip Plant Monster," 15 February 1979, p. 4A.

42. Committee for the Rational Use of Our Forests, Inc., "Wood Energy in Michigan: An Analysis of Impact and Alternatives to the Proposed Generating Plant at Hersey, Michigan," mimeo, Hersey, Mich., 26 March 1979, 21 p.

43. *Osceola County Herald*, "The Wood Chip Controversy: Pro: Money's A Key . . . Con: Use is Inefficient," 12 April 1979, pp. 3, 8B.

44. Ibid.

45. *Osceola County Herald*, "DNR Delays Wood Decision," 13 September 1979, p. 3A.

46. *Osceola County Herald*, "Wood Chip Plant Fuel Supply Approved for 10 Years," 18 October 1979, p. 1A.

47. *Osceola County Herald*, "CRUF Reaffirms Stand Opposing Hersey Wood Chip Plant," 25 October 1979, p. 1A.

48. Hersey's Solid Waste and Toxic and Hazardous Substances Disposal Ordinance is the ordinance in question here.

49. *Osceola County Herald*, "Scrambling to the Need," 29 November 1979, p. 4A.

50. *Osceola County Herald*, "What's Really Being Planned for the Hersey Area?" 18 October 1979, p. 4A. See also Daverman and Associates, Inc., *A Feasibility Study for a Waste-Wood Electric Generating Plant*, Section 3.

51. *Osceola County Herald*, "Commissioners Urged To Reconsider," 8 November 1979, p. 4A.

52. Daverman and Associates, Inc., *A Feasibility Study for a Waste-Wood Electric Generating Plant*.

53. *Osceola County Herald*, "Board Commended," 27 December 1979, p. 4A.

54. The lead article in the first issue of the newspaper in 1980 reviewed the top stories in Osceola County during 1979. The Hersey WFPP controversy was ranked third.

55. *Osceola County Herald*, "RDF Hot Issue in Hersey," 17 April 1980, pp. 1, 4B.

56. *Osceola County Herald*, "Your Choice: Fact or Fiction," 24 April 1980, p. 4A.

57. *Detroit News*, "Tiny Town Beats Power Plant Bid by Two Utilities," 18 May 1980, p. 5B.

58. Committee for the Rational Use of Our Forests, Inc., "An Update on the Hersey Wood Fired Power Plant," mimeo, Hersey, Mich., 1 June 1980, 2 p.

59. *Grand Rapids Press*, "Wood-Fueled Power Plant Is Put on Shelf," 11 September 1980, p. 12A.

Chapter 5. Subsequent Wood Energy Controversies in the United States

1. *Grand Rapids Press*, "Many Towns Want Wood Electric Plant," 25 May 1980, p. 13A.

2. *Osceola County Herald*, "Local Towns Offer Sites for Plant," 19 June 1980, p. 1A.

3. *Osceola County Herald*, "CPC to Delay Wood Plant Study," 18 September 1980, p. 1A.

4. *Grand Rapids Press*, "Ogemaw Beckons Waste-Fueled Power Plant," 27 June 1980, p. 1A.

5. CRUF, "An Update on the Hersey Wood Fired Power Plant."

6. *Big Rapids Pioneer*, "Farwell To Host Wood Meeting," 17 June 1980, p. 1A.

7. *Grand Rapids Press*, "Wood-fueled Power Plant Idea Is Debated," 3 August 1980, p. 13A.

8. *Grand Rapids Press*, "Wood-Fueled Power Plant Is Put on Shelf"; *Lansing State Journal*, "Electricity from Wood Must Wait," 12 September 1980, p. 8B.

9. An editorial in the *Detroit News* ("Wood as Energy," 28 January 1983, p. 12A) favoring large-scale wood energy development used as its main examples both the newly completed Dow Corning SECO plant and the receipt of a grant by Central Michigan University to build a wood chip-burning facility.

10. This description of the Midland SECO facility is taken from the following accounts: *Detroit News*, "Cash in the Chips: Firm Sees Big Savings on Energy in New Wood-Fueled Power Plant," 18 January 1983, p. 4B; *Detroit Free Press*, "Industrial-Strength Wood Power Is Back," 2 March 1984, p. 1C; *Grand Rapids Press*, "Wood-Chip Fuel Gets a Warm Reception at Dow Corning's New Plant in Midland," 18 January 1983, p. 7A; *Lansing State Journal*, "Plant Powered by Wood," 18 January 1983, p. 2B.

11. G. Graff, "Wood-fueled Cogeneration: Keeping a Good Thing Going in Michigan," *Northern Logger and Timber Processor* 37 (November 1988): 26.

12. Dow Corning has received national attention for its innovative cogeneration project, e.g., J. Makansi, "Wood-Burning Cogen System Makes Debut in the Chemical Industry," *Power* 128 (February 1984): 103–4; H. Rolka and P. E. Sworden, "SECO—Dow Corning's Wood Fired Industrial Cogeneration Project," in *Advances in Energy Productivity: Proceedings of the 5th World Energy*

Congress, September 14–17, 1982 (Atlanta: Association of Energy Engineers, 1982), pp. 147–50; *Pollution Engineering*, "Woodburning Cogeneration Power Plant Completed on Time, Near Budget," 16 (April 1984): 6; *Plant Engineering*, "Woodburning Cogeneration System Cuts Plant Energy Costs," 38 (9 February 1984): 26+.

13. This summary of the Central Michigan University wood burning facility is taken from articles that appeared in the school newspaper—the *CMU Centralight*, literature from the university, and G. Ponczak, "University Converts Heat Plant in Order To Use Local Timber," *Energy User News*, 11 February 1985, pp. 2–3. A second wood energy conversion project in the planning stage as of 1985 was the installation of a steam turbine to produce part of the university's electricity.

14. By 1989 the CMU plant had ceased to be a model of wood energy development. In a personal note dated June 1989, a facilities management official for the university outlined plans to curtail the wood energy system due to high maintenance and operations costs. A 3–MW gas turbine system with a waste heat boiler was the planned replacement.

15. *Mining Journal*, "Consumers' Woes Hurting Small Power Plants," 9 April 1985, p. 5A.

16. *Detroit News*, "Power Play: Firms Drop Utilities, Create Their Own Energy," 24 January 1985, pp. 1–2A.

17. *Lansing State Journal*, "North Michigan Town Finds Key To Revitalization," 16 November 1986, p. 7B.

18. D. Hacker, "Three Villages Branch Out by Planning Wood Power," *Detroit Free Press*, 13 October 1987, pp. 3, 10A.

19. Graff, "Wood-fueled Cogeneration: Keeping a Good Thing Going in Michigan," p. 27.

20. See *Northern Logger and Timber Processor*, "Vermont's Burlington Electric Under Fire," 34 (January 1986): 4; *Biologue*, "Burlington Electric Departments Problems Continue," 3, 2 (1986): 2, 18.

21. This account of the Indian River controversy is largely derived from the following newspaper accounts: *Detroit News*, "Electric Plant Bid Sets Off Sparks," 15 March 1984, pp. 1, 2F; *Lansing State Journal*, "Residents Split on Energy Plant," 18 March 1984, p. 7B; *Grand Rapids Press*, "Judge Halts Plans for Cheboygan Wood-Burning Plant," 23 November 1984, p. 13A.

22. T. Blackman, "Mechanized Thinning Helps Forest, Power Production," *Forest Industries* 112, 10 (1986): 24–25; T. Blackman, "Another Wood-fueled Plant Fires Up Its Generators," *Forest Industries* 113, 10 (1986): 23.

23. J. Applegate, "Ultrasystems Pins Hopes on Wood Power," *Los Angeles Times*, sec. IV, pp. 1, 4.

24. This section is largely a summary of the study presented by Fortmann and Starrs ("Burning Issues: Power Plants and Rights Over Resources") at the First National Symposium on Social Science in Natural Resource Management, Corvallis, Oregon, May 1986. It was published as L. Fortmann and P. Starrs, "Power Plants and Resource Rights" in R. G. Lee, D. R. Field, and W, R. Burch, Jr. (eds.) *Community and Forestry: Continuities in the Sociology of Natural Resources* (Boulder, CO: Westview Press, 1990), pp. 179–193. They used pseudonyms in their reports but actual names and places are included in this summary.

25. Fortmann and Starrs, "Power Plants and Resource Rights," p. 182.

26. Ibid.

27. Ibid., pp. 187–189.

28. Ibid., p. 187.

29. Fortmann and Starrs, "Burning Issues! Power Plants and Rights over Resources," p. 16.

30. Ibid., p. 18.

31. Fortmann and Starrs, "Power Plants and Resource Rights," p. 180.

32. Ibid., p. 191.

33. Ibid., p. 189.

34. Ibid., p. 191.

35. Ibid., pp. 187–189.

36. As an example of the ideology of WFPP opponents, see Committee for Rational Use of Our Forests, Inc., "Wood Energy as a 'Soft Technology': An Alternative Proposal," mimeo, Hersey, Michigan, 27 March 1979, 2 p.

37. Perhaps the most important variable in either study, the population migration turnaround, is given short shrift by Fortmann and Starrs. For example, they state at the outset (Fortmann and Starrs, "Power Plants and Resource Rights," p. 181) that "...urban refugees, alone, do not a protest make...Explanation of their [Westwood and Quincy] differing reactions lies elsewhere." The present evaluation suggests that they too quickly dismissed the theoretical importance of the population migration turnaround. The work of Sokolow ("Local Politics and the Turnaround Migration: Newcomer-Oldtimers Relations in Small Communities"), another California scholar, and Blahna, who published a paper ("Social Bases for Resource Conflicts in Areas of Reverse Migration") in the same volume as Fortmann and Starrs, are noticably uncited. The Michigan cases of WFPP development conflict suggest that the population migration turnaround is far more fundamental to the explanation of natural resource conflict in rural areas than the gross structural variables embraced in the California study.

38. Harris, "Vermont: Goodbye Coal, Hello Trees."

39. Osceola County Herald, "Vermont Wood Chip Plant Shelved," 15 November 1979, p. 1A.

40. This description is based on articles that appeared in the Burlington Free Press, July–October 1979.

41. K. Anderson, "Burlington's Wood and Water," Public Power 39, 6 (1981): 41–4.

42. New York Times, "Power Plant in Vermont Fueled by Wood Chips," 16 April 1984, p. 14A.

43. D. F. Dennis and S. J. Dresser, "Burlington's Wood-Burning Utility Company," Journal of Forestry 83, 2 (1985): 102.

44. Dennis and Dresser, "Burlington's Wood-Burning Utility Company," p. 102; J. Reason, "Wood-fired Powerplants Are Not Without Problems," Power 130 (August 1986): 66–67.

45. Dennis and Dresser, "Burlington's Wood-Burning Utility Company," p. 102.

46. This account is taken from Northern Logger and Timber Processor, "Vermont's Burlington Electric under Fire," Biologue, "Burlington Electric Departments Problems Continue," and B. Gove, "The Saga of Burlington Electric: The Current Status of What Was To Be a Model Wood Energy System," Northern Logger and Timber Processor 35 (November 1986): 26–30.

47. Biologue, "Wood Fired Plants Facing Opposition," 3, 2 (1986): 1; B. Gove, "The Saga of Burlington Electric," p. 27.

48. Reason, "Wood-fired Powerplants Are Not without Problems."

49. D. C. Rinebolt, "Wood Power for the Future," Public Power 46, 6 (1988): 37–39.

50. B. Gove, "Wood-fueled Power Plants Promise Uncertain Future for Vermont/New Hampshire," *Northern Logger and Timber Processor* 36 (November 1987): 22.

51. E. Johnson, "Biomass Energy: Caught in the Middle of New Hampshire," *Northern Logger and Timber Processor* 37 (November 1988): 6–7.

52. Ibid., p. 7.

53. Gove, "Wood-fueled Power Plants Promise Uncertain Future for Vermont/New Hampshire," p. 23.

54. *Northern Logger and Timber Processor*, "More than 20 Biomass Energy Projects Planned for Maine," 33 (January 1985): 8.

55. *Northern Logger and Timber Processor*, "Despite Oil Price Plunge, Wood Energy Market Looks Strong—For the Moment at Least," 34 (March 1986): 1.

56. *Northern Logger and Timber Processor*, "More than 20 Biomass Energy Projects Planned for Maine," p. 8.

57. *Northern Logger and Timber Processor*, "New England Wood Energy Situation Brightens Up," 35 (January 1987): 1.

58. *Northern Logger and Timber Processor*, "Environmentalist Groups Sue Over Proposed Maine Biomass Plant," 36 (November 1987): 2.

59. *Northern Logger and Timber Processor*, "Wood-fired Power Plants in Maine's Aroostook County To Be Built," 36 (June 1988): 2.

60. *Northern Logger and Timber Processor*, "Town Planning Commission Approves Biomass-Burning Plant in Ryegate, VT," 35 (February 1987): 2.

61. *Northern Logger and Timber Processor*, "Upstate New York Plant Proposals under Scrutiny," 33 (January 1985): 8.

62. Gove, "Wood-fueled Power Plants Promise Uncertain Future for Vermont/New Hampshire," pp. 22–23.

63. *Northern Logger and Timber Processor*, "Ultrapower Hits Yet Another Potential Snag in New York State," 34 (October 1985): 2.

64. *Northern Logger and Timber Processor*, "Ultrapower Blames Plant Uncertainty on NY Public Service Commission," 35 (November 1986): 1.

65. *Northern Logger and Timber Processor*, "New York Community Forges Ahead with Its Own Wood-Burning Plans," 36 (June 1988): 1.

Chapter 6. Policy Implications for Large-Scale Wood Energy Development

1. Nelkin, *Nuclear Power and Its Critics*, chapter 1.

2. Charles Hewett, a wood-energy researcher in the Resource Policy Center at Dartmouth University, calls for wood-energy development at much smaller scale. He would limit plant size to 3–5 MW in order to match decentralized wood sources. His suggestion is cited in E. P. Frank, "The Yankee Forest: Will It Be Plundered or Preserved by Wood Heat?" *New Roots*, no. 9 (January–February 1980): 29–32.

3. For a discussion of this point see J. P. Holdren, G. Morris, and I. Mintzer, "Environmental Aspects of Renewable Energy Sources," *Annual Review of Energy* 5 (1980): 241–91.

4. M. D. Moore (ed.), *Proceedings of Governor James J. Blanchard's Conference on Forest Resources: Creating 50,000 New Jobs in Michigan Forest Products Industries, March 22–23, 1983, Michigan State University, East Lansing, MI* (Lansing, Mich.: State of Michigan, 1983).

5. Ibid., p. 70.

6. L. F. Lokken, *Michigan's Forest Products Industry Development Program: A Progress Report* (Lansing, Mich.: Office of the Governor, May 1985).

7. Michigan, State of, *Michigan Wood Energy Development Plan*, p. 1.

8. Ibid, p. iii.

9. Ibid, p. 15.

10. Ibid, p. 1.

11. Ibid.

12. *Lansing State Journal*, "Loggers, Environmentalists Fight Forest Plan," 12 December 1986, p. 3B; *Lansing State Journal*, "New Wood Processes Step Up Timber Needs, Spark Battle over Land," 19 October 1986, p. 1E.

13. Hewett and Peterson, "The Forest Resource," p. 15.

14. Hacker, "Three Villages Branch Out by Planning Wood Power."

Chapter 7. Values, Facts, and Ideology in the Hersey Controversy

1. L. H. Tribe, C. S. Schelling, and J. Voss (eds.), *When Values Conflict: Essays on Environmental Analysis, Discourse, and Decision* (Cambridge, Mass.: Ballinger Publishing Co., 1976); E. L. Hayman, and B. Stiftel, *Combining Facts and Values in EIA* (Boulder, Colo.: Westview Press, 1988).

2. R. M. Williams Jr., *American Society*, 3rd ed. (New York: Knopf, 1970).

3. Ravetz, "Scientific Knowledge and Expert Advice in Debates about Large Technological Innovations." S. G. Hadden ("Technical Information for Citizen Participation," *Journal of Applied Behavioral Science* 7, 4 (1981): 546) points out that technical information alone cannot settle value questions such as those involved in accepting risks and determining the tradeoffs between risks and benefits.

4. Mulkay, *Science and the Sociology of Knowledge*, p. 120.

5. Barbour, *Technology, Environment, and Human Values*, pp. 126–28.

6. Bartlett (*The Reserve Mining Controversy*, p. 78) similarly discovered that science was used to support positions taken for nonscientific reasons.

7. Nelkin, "Science and the Polity," p. 7.

8. See, for example, the case of the Santa Barbara oil spill as documented by Molotch, "Oil in Santa Barbara and Power in America."

9. Ibid, p. 140.

10. Daverman and Associates, Inc., *A Feasibility Study for a Waste-Wood Electric Generating Plant*, Appendix I.

11. Table 4 constitutes the "general conclusions ... regarding the attitudes prevailing at all three [prime site] meetings" from the feasibility study.

12. Daverman and Associates, Inc., *A Feasibility Study for a Waste-Wood Electric Generating Plant*. Documents researched, written, and disseminated by the Committee for Rational Use of Our Forests include: "Wood Energy in Michigan: An Alternative Proposal," mimeo, Hersey, Mich., 27 March 1979, 2 p.; "Protect Your Wood Supply," pamphlet, Hersey, Mich., February 1980; "A Second Look at Consumers Power Company's Proposed Hersey Wood-Fired Power Plant," mimeo, Hersey, Mich., March 1980, 7 p.; and, "An Update on the Hersey Wood Fired Power Plant."

13. A. B. Motz, "Toward Assessing Social Impacts: The Diachronic Analysis of Newspaper Contents," in K. Finsterbusch and C. P. Wolf (eds.), *Methodology of Social Impact Assessment* (Stroudsburg, Pa.: Dowden, Hutchinson, and Ross, 1977), pp. 303–13.

14. M. Altimore, "The Social Construction of a Scientific Controversy: Comments on Press Coverage of the Recombinant DNA Debate," *Science, Technology, and Human Values* 7 (Fall 1982): 24–31.

15. See Frankena, "Rethinking the Scale of Biomass Energy Facilities."

16. Sells, "Use of Wood as an Energy Source"; CRUF, "Wood Energy as a 'Soft Technology'."

17. Daverman and Associates, Inc., *A Feasibility Study for a Waste-Wood Electric Generating Plant*, Appendix I, p. 108.

18. Ibid., Appendix I, pp. 109–110.

19. CRUF, "A Second Look at Consumers Power Company's Proposed Hersey Wood-Fired Power Plant."

20. See "Issues in the Hersey Controversy" in chapter 4, p. 119.

21. Cited in B. Miller, "Resourceful People," *Michigan Natural Resources* 48, 5 (September–October 1979): 45.

Chapter 8. The Politics of Expertise in the Hersey Controversy

1. Tichenor, Donohue, and Olien, *Community Conflict and the Press*, p. 205.
2. Sells, "Use of Wood as an Energy Source," p. 91.
3. Tichenor, Donohue, and Olien, *Community Conflict and the Press*.
4. Daverman and Associates, Inc., *A Feasibility Study for a Waste-Wood Electric Generating Plant*.
5. CRUF, "Wood Energy in Michigan."
6. *Osceola County Herald*, "Your Choice: Fact or Fiction," 24 April 1980, p. 4.
7. CRUF, "Wood Energy in Michigan."
8. *Osceola County Herald*, "Fight Facts Were Presented," 8 May 1980, p. 2A.
9. *Osceola County Herald*, "The Demise of the Hersey Woodburner," 31 July 1980, p. 5B.
10. Ibid.
11. *Osceola County Herald*, "CRUF Is Not Short on Facts," 5 June 1980, p. 2A.
12. Nelkin, "The Political Impact of Technical Expertise," pp. 51–54.
13. Hays, *Beauty, Health, and Permanence*, pp. 394, 330.
14. Hays, "The Structure of Environmental Politics since World War II," p. 727.
15. Feldman and Milch, *Technocracy Versus Democracy*, 146; Sclove, "Decision-Making in a Democracy," p. 46.
16. As Bartlett (*The Reserve Mining Controversy*, p. 226) discovered, scientists were expected to take sides and did just that. Rayner and Peerla "(To Spray or Not to Spray?" p. 28) discuss the predicament of the expert, that is, he or she can only apply technique to externally determined goals.
17. Nowotny, "Experts and Their Expertise," p. 236.
18. Edwards and von Winterfeldt, "Public Disputes about Risky Technologies," p. 87.
19. Goodman, "Ecological Expertise," pp. 354–155.

Chapter 9. Understanding Controversy: The Case of Wood Energy Development

1. Hays, "Value Premises for Planning and Public Policy," p. 149.

2. Hays, "Value Premises for Planning and Public Policy." This thesis is more fully explored in Hays, *Beauty, Health, and Permanence*.

3. Aldrich (*Communities of Opposition*) characterizes siting as destructive of community values. The strategy of exploiting areas of uncertainty is discussed in Nelkin, "Science, Technology, and Political Conflict."

4. Nichols, "The De-Institutionalization of Technical Expertise," p. 38.

5. See Nelkin, "Science, Technology, and Political Conflict."

6. Mazur (*The Dynamics of Technical Controversy*, p. 17) notes that the most common rhetorical device observed in controversies is "There is no evidence to show that . . . ."

7. Nelkin, "The Political Impact of Technical Expertise." The reader may wish to compare the chronology in Appendix I with tables 18 and 19.

8. Nelkin, "The Political Impact of Technical Expertise," p. 42.

9. Tichenor, Donohue, and Olien (*Community Conflict and the Press*) propose "phases of conflict" that are nearly congruent with the Mazur scheme, but use different terminology and focus more on the role of communication. The stages identified are 1) initiation, 2) conflict definition, 3) public phase, 4) one or more legitimation phases, and 5) other phases depending upon the nature, extent, and organized basis for the conflict. The public phase requires recognition of the issue by different segments of the community and generally involves secondary communication, e.g., via the newspaper. By this criterion the Hersey controversy reached the public phase very quickly. The conflict definition phase was almost instantaneous, a function perhaps of the barrage of public hearings in September–October 1978.

10. Nelkin, "The Political Impact of Technical Expertise," pp. 43–44.

11. December 1979–February 1980, refer to Appendix I.

12. Nelkin, "The Political Impact of Technical Expertise," pp. 44–45. This rhetorical license has been observed in numerous other controversies. For example, Bartlett (*The Reserve Mining Controversy*, p. 224) found that: "Allegations of bias, conflict of interest, bad faith, cover-up, suppression, and improper relationships abounded. Reserve [a mining company] charged that it was being convicted in the kangaroo court of public opinion by the use of innuendo, bad science, and unsubstantiated conclusions 'craftily written to sound bad.' Environmentalists charged Reserve with defending itself by impugning the motives and integrity of government scientists and officials, and by cleverly misleading the lay public with its presentations and press releases."

13. The rhetoric of this debate is apparent in the captions for the letters, e.g., "Your Choice: Fact or Fiction," "Somewhat Distorted Reasoning," "More CRUF Fiction," "No Shred of Democracy Left," "Cause of Accuracy Not Well Served," and "Here We Go Again."

14. Nelkin, "The Political Impact of Technical Expertise," p. 48.

15. Casper and Wellstone, *Powerline*.

16. Nichols, "The De-Institutionalization of Technical Expertise," p. 46.

Chapter 10. The Challenges of Technocratic and Democratic Decision Making

1. Coser, "Conflict: Social Aspects," p. 235.

2. Klessing and Strite, *The ELF Odyssey*, p. 234.

3. Nichols, "The De-Institutionalization of Technical Expertise," p. 38.

4. The history of the changing scale of activities and the drive for large-scale

systems is treated in Hays, *Beauty, Health, and Permanence*. For a summary, see Hays, "Value Premises for Planning and Public Policy."

5. See Blahna, "Rural Population Growth and Social Information Needs of Forest Management Agencies."

6. For two excellent accounts of the dilemma of technocratic versus democratic decision making, see DeSario and Langton, "Citizen Participation and Technocracy," and Nelkin, "Scientific Knowledge, Public Policy, and Democracy."

7. Aldrich, *Communities of Opposition*, p. 40.

8. DeSario and Langton, "Citizen Participation and Technocracy," pp. 5–7.

9. Aldrich, *Communities of Opposition*, p. i.

10. Hays, "Value Premises for Planning and Public Policy," p. 149.

11. President's Commission for a National Agenda for the Eighties, *Science and Technology: Promises and Dangers in the Eighties* (Washington, D.C.: U.S. Government Printing Office, 1981), p. 3.

12. Readers who wish to further explore the field of citizen participation in connection with environmental and natural resource issues should consult F. Frankena and J. K. Frankena, *Citizen Participation in Environmental Affairs, 1970–1986: A Bibliography*, Studies in Social History no. 8 (New York: AMS Press, 1988).

13. The quote is from DeSario and Langton, "Citizen Participation and Technocracy," p. 13. Nelkin and Pollak ("Problems and Procedures in the Regulation of Technological Risk," p. 247) suggest the following conditions for resolving the tension between technocracy and democracy: ". . . a careful definition of the agenda that gives due weight to social and political concerns, the appropriate involvement of affected interests, an unbiased management, a fair distribution of expertise, and a real margin of choice."

Bibliography

Books, Reports, and Journal Articles

Albrecht, S. L. "Community Response to Large Scale Federal Project: The Case of the MX." In S. H. Murdock et al. (eds.), *Nuclear Waste: Socioeconomic Dimensions of Long-term Storage*. Boulder, Colo.: Westview Press, 1983, pp. 233–50.

———. "Socio-Cultural Factors and Energy Resource Development in Rural Areas of the West." *Journal of Environmental Management* 7, 1 (1978): 73–90.

Aldrich, B. C. *Communities of Opposition: Energy Facility Siting in Minnesota*. Winona, Minn.: Winona State University, 1980.

Altimore, M. "The Social Construction of a Scientific Controversy: Comments on Press Coverage of the Recombinant DNA Debate." *Science, Technology, and Human Values* 7 (Fall 1982): 24–31.

American Society of Planning Officials. *Subdividing Rural America: Impacts of Recreational Lot and Second Home Development*. Prepared for the Council on Environmental Quality. Washington, D.C.: Government Printing Office, 1976.

Barbour, I. G. *Technology, Environment, and Human Values*. New York: Praeger, 1980.

Bartlett, R. V. *The Reserve Mining Controversy: Science, Technology, and Environmental Quality*. Bloomington, Ind.: Indiana University Press, 1980.

Beale, C. L. "The Reserve Shift of U.S. Population to Nonmetropolitan Areas, 1970–75." *International Regional Science Review* 2, 2 (1975): 113–22.

———. *The Revival of Population Growth in Nonmetropolitan America*. Washington, D.C.: Economic Research Service, USDA, 1975.

Benveniste, G. *The Politics of Expertise*. 2nd ed. Boston, Mass.: Boyd and Fraser Pub. Co., 1977.

Blahna, D. J. "Rural Population Growth and Social Information Needs of Forest Management Agencies." In L. N. Wenner (eds.), *Social Science Information and Resource Management: Proceedings from an Interagency Symposium*. Washington, D.C.: U.S. Forest Service, 1985, pp. 23–31.

Blahna, D. J. "Social Bases for Resource Conflicts in Area of Reverse Migration." In R. G. Lee, D. R. Field, and W. R. Burch, J., eds. *Community and Forestry: Continuities in the Sociology of Natural Resources*. Boulder, CO: Westview Press, 1990, pp. 159–78.

Burwell, C. C. "Solar Biomass Energy: An Overview of U. S. Potential." *Science* 199 (10 March 1978): 1041–48.

Caine, T. A. "The Willmar Bank Strike." In R. P. Wolensky and E. J. Miller (eds.), *The Small City and Regional Community: Proceedings of the Conference on the Small City and Regional Community.* Vol. 2. Stevens Point, Wis.: Foundation Press, Inc., 1979, pp. 158–66.

Caldwell, L. K., Hayes, R. and I. M. MacWhirter. *Citizens and the Environment: Case Studies in Population Action.* Bloomington: Indiana University Press, 1976.

Casper, B. M. and P. D. Wellstone. *Powerline: The First Battle of America's Energy War.* Amherst, Mass.: University of Massachusetts Press, 1981.

Catton, W. R., Jr. "Depending on Ghosts." *Humboldt Journal of Social Relations* 2 (Fall/Winter 1974): 45–49.

Chen, E. *PBB: An American Tragedy.* Englewood Cliffs, N.J.: Prentice-Hall, 1979.

Christenson, J. A. "Value Configurations for Ruralites and Urbanites: A Comment on Bealer's Paper." *Rural Sociologist* 1, 1 (January 1981): 42–47.

Clark, I. D. "Expert Advice in the Controversy about Supersonic Transport in the United States." *Minerva* 12 (October 1974): 416–32.

Cluett, C., Greene, M., and L. Radford. *Individual and Community Response to Energy Facility Siting: An Annotated Biliography.* Public Administration Series no. 493. Vance Bibliographies, Monticello, Ill., May 1980.

Cluett C., Greene, M., and L. Radford. *Individual and Community Response to Energy Facility Siting: A Review of the Literature.* B–HARC–411–045. Battelle Affairs Research Centers, Seattle, Wash., November 1979.

Committee for the Rational Use of Our Forests, Inc. "Protect Your Wood Supply." Pamphlet. Hersey, Mich., February 1980.

————. "A Second Look at Consumers Power Company's Proposed Hersey Wood-Fired Power Plant." Mimeo. Hersey, Mich., March 1980. 7 p.

————. "An Update on the Hersey Wood Fired Power Plant." Mimeo. Hersey, Mich., 1 June 1980. 2 p.

————. "Wood Energy as a 'Soft Technology': An Alternative Proposal." Mimeo. Hersey, Mich., 27 March 1979. 2 p.

————. "Wood Energy in Michigan: An Analysis of Impact and Alternatives to the Proposed Generating Plant at Hersey, Michigan." Mimeo. Hersey, Mich., 26 March 1979. 21 p.

Cooper, J. A. "Environmental Impact of Residential Wood Combustion Emissions and Its Impact." *Journal of the Air Pollution Control Association* 30 (1980): 855–61.

Cortese, C. "Rapid Growth and Social Change in Western Communities." *Social Impact Assessment* 40/41 (April–May 1979): 1–7.

Coser. L. A. "Conflict: Social Aspects." *International Encyclopedia of the Social Sciences.* New York: Macmillan Co. and the Free Press, 1968, pp. 232–36.

————. *The Functions of Social Conflict.* Glencoe, Ill.: The Free Press, 1956.

Coyer, B. W. and D. S. Schwerin. "Bureaucratic Regulation and Farmer Protest in the Michigan PBB Contamination Case." *Rural Sociology* 46, 4 (Winter 1981): 703–23.

Crain R. L., Katz, E., and D., B. Rosenthal. *The Politics of Community Conflict: The Fluoridation Decision.* New York: Bobbs-Merrill, 1969.

Daverman and Associates. *A Feasibility Study for a Waste-Wood Electric Generating Plant.* Prepared for Consumers Power Company, Morbark Industries, Inc. and Wolverine Electric Cooperative, Grand Rapids, Mich., February 1979.

DeSario, J. and S. Langton. "Citizen Participation and Technocracy." In J.

DeSario and S. Langton (eds.), *Citizen Participation in Public Decision Making.* New York: Greenwood Press, 1987, pp. 3–17.

Dennis, D. F. and S. J. Dresser, "Burlington's Wood-Burning Utility Company." *Journal of Forestry* 83, 2 (1985): 101–4.

Edwards, W. and D. von Winterfeldt. "Public Disputes about Risky Technologies: Stakeholders and Arenas." In V. Covello, J. Menkes, and J. Mumpower (eds.), *Risk Evaluation and Management.* New York: Plenum Press, 1984, pp. 69–92.

Egginton, J. *The Poisoning of Michigan.* New York: W. W. Norton and Co., 1980.

Enbar, M. "Equity in the Social Sciences." In R. E. Kasperson (ed.), *Equity Issues in Radioactive Waste Management.* Cambridge, Mass.: Oelgeschlager, Gunn and Hain, 1983, pp. 3–23.

Farkas, E. J. "The NIMBY Syndrome." *Alternatives* 19, 2–3, (1982): 47–50.

Feldman, E. J. and J. Milch. *Technocracy versus Democracy: The Comparative Politics of International Airports.* Boston, Mass.: Auburn House Pub. Co., 1982.

Fortmann L. and P. Starrs. "Burning Issues: Power Plants and Rights over Resources." Paper presented at the First National Symposium on Social Science in Natural Resource Management, Corvallis, Oreg., May 1986.

———. "Power Plants and Resource Rights." In R. G. Lee, D. R. Field, and W. R. Burch, Jr. (eds.), *Community and Forestry: Continuities in the Sociology of Natural Resources.* Boulder, CO: Westview Press, 1990, pp. 179–93.

Frankena, F. *Community Impacts of Rapid Growth in Nonmetropolitan Areas: A Literature Survey.* Rural Sociology Series no. 9. Michigan Agricultural Experimental Station, East Lansing, Mich., June 1980.

———. "Defeat of a Wood-Fired Electric Power Plant: A Study in Social Change." *Society and Natural Resources: An International Journal* 1, 2 (1988): 167–83.

———. "The Emergent Social Role and Political Impact of the Voluntary Technical Expert." *Environmental Impact Assessment Review* 8, 1 (1988): 73–84.

———. *Experts and Expertise in Environmental Litigation: A Bibliography.* Public Administration Series no. P-1909. Vance Bibliographies, Monticello, Ill. April 1986.

———. "Facts, Values, and Technical Expertise in a Renewable Energy Siting Dispute." *Journal of Economic Psychology* 4 (1983): 131–47.

———. *The Impact of Technical Expertise in a Nonmetropolitan Siting Dispute: A Case Study of the Hersey Wood-Fired Power Plant Controversy.* Ph. D. dissertation. Michigan State University, East Lansing, Mich., 1982 (available from University Microfilms International, Ann Arbor, Mich.).

———. "Large-Scale Wood Energy Development in Michigan: An Assessment of Citizen Involvement." *Renewable Resources Journal* 2, 1 (1989): 14–20.

———. "Regional Socioeconomic Impacts of Declining Net Energy. *"Urban Ecology* 3, 2 (1978): 101–10.

———. "Rethinking the Scale of Biomass Energy Conversion Facilities: The Case of Wood-Electric Power." *Biomass: An International Journal* 14, 4 (1987): 149–71.

———. "Wood-fired Power Plants: Public Controversies Reveal Major Social Changes." *Journal of Forestry* 87, 4 (1989): 18–23.

———. and J. K. Frankena. *Citizen Participation in Environmental Affairs, 1970–1986: A Bibliography.* Studies in Social History no. 8. New York: AMS Press, 1988.

———. and T. Koebernick. "The Pattern of Recent Housing Growth in a

Nonmetropolitan County: Effects of Environment and Location." *Growth and Change* 15, 4 (1984): 32–42.

Fuguitt, G. V. and C. L. Beale. "Post-1970 Shifts in the Pattern of Population Change in the North Central Region." In J. A. Beegle and R. L. McNamara (eds.), *Patterns of Migration and Population Change in America's Heartland.* East Lansing, Mich.: Michigan Agricultural Experiment Station, 1978, pp, 14–21.

Gatrell, J. W. "Community as a Social Collective." In G. F. Summers and A. Selvik (eds.), *Energy Resource Communities.* Madison, Wis.: MJM Publishing Co., 1982, pp. 199–218.

Georgescu-Roegen, N. *The Entropy Law and the Economic Process.* Cambridge, Mass.: Harvard University Press, 1971.

Gladwin, T. N. "Patterns of Environmental Conflict over Industrial Facilities in the United States, 1970–80." *Natural Resources Journal* 20, 2 (1980): 243–74.

Goodman, D. "Ecological Expertise." In H. A. Feiveson et al. (eds.), *Boundaries of Analysis: An Inquiry into the Tocks Island Dam Controversy.* Cambridge, Mass.: Ballinger, 1976, pp. 317–60.

Grace, A. *Biomass: Solar Energy from Farms and Forests.* Prepared for the Solar Energy Research Institute, U.S. Government Printing Office, Washington, D.C., 1980.

Gross, A. G. "Public Debates as Failed Social Dramas: The Recombinant DNA Controversy." *Quarterly Journal of Speech* 70, 4 (1984): 397–409.

Hadden, S. G. "Technical Information for Citizen Participation." *Journal of Applied Behavioral Science* 7, 4 (1981): 537–49.

Harthill, M. (ed.). *Hazardous Waste Management: In Whose Backyard?* AAAS Selected Symposium Series. Boulder, Colo.: Westview Press, 1984.

Hays, S. P. *Beauty, Health, and Permanence; Environmental Politics in the United States, 1955–1985.* New York: Cambridge University Press, 1987.

———. "The Structure of Environmental Politics since World War II." *Journal of Social History* 14, 4 (Summer 1981): 719–38.

———. "Value Premises for Planning and Public Policy: The Historical Context." In R. N. L. Andrews (ed.), *Land in America; Commodity or Natural Resource?* Lexington, Mass.: Lexington Books, 1979, pp. 149–66.

Henderson, H, *Creating Alternative Futures: The End of Economics.* New York: Berkley Publishing Corp., 1978.

Hendricks, J. W. "Public Participation and Democratic Decision Making on Energy Issues." *Social Science Energy Review* 1, 2 (1978): 1–25.

Herbers, J. *The New Heartland.* New York: Times Books, 1986.

Hewett, C. K. and E. Peterson. "The Forest Resource: Emerging Conflicts and the Need for Action." In C. E. Hewett and T. E. Hamilton (eds.), *Forests in Demand: Conflicts and Solutions.* Boston, Mass.: Auburn House Pub., 1982, pp. 13–28.

———. et al. "Wood Energy in the United States." *Annual Review of Energy* 6 (1981): 139–70.

Hiser, M. L. (ed.). *Wood Energy: Proceedings of Governor William G. Milliken's Conference, November 29, 1977, Ann Arbor, Michigan.* Ann Arbor, Mich.: Ann Arbor Science Publishers, 1978.

Hiser, M., MacGregor, L., Padgett, J., Rudd, J., and T. Heck, *Wood Fueled Power Generation: A Potential Source of Energy for Northern Michigan.* Michigan Public Service Commission, Lansing, November 1977.

Holdren, J. P. "The Nuclear Controversy and the Limitations of Decision-making by Experts." *Bulletin of the Atomic Scientists* 32 (March 1976): 20–22.

———. Morris, G., and I. Mintzer, "Environmental Aspects of Renewable Energy Sources." *Annual Review of Energy* 5 (1980): 241–91.

Hoos, I. R. "The Credibility Issue." In *Essays on Issues Relevant to the Regulation of Radioactive Waste Management.* NUREG–0412. Office of Nuclear Material Safety and Safeguards, U.S. Nuclear Regulatory Commission, May 1978, pp. 21–30.

Horowitz, I. L. "Sociological and Ideological Conceptions of Industrial Development." *American Journal of Economics and Sociology* 23 (October 1964): 351–74.

Huber, T. P., Gruntfest, E. C., and L. Lapalme-Roy. "The Use of Wood as Fuel in North America: Prospects and Problems." *Journal of Environmental Systems* 14, 4 (1984–85): 321–32.

Hyman, E. L. and B. Stiftel. *Combining Facts and Values in EIA.* Boulder, Colo.: Westview Press, 1988.

Institute for Environmental Negotiation (ed.). *Not-in-My-Backyard!: Community Reaction to Locally Unwanted Land Use.* Charlottesville, Va., 1985.

Jakimo, A. and I. C. Bupp. "Nuclear Waste Disposal: Not in My Backyard." *Technology Review* 80, 5 (1978): 64–72.

Johnston, R. "Controlling Technology: An Issue for the Social Studies of Science." *Social Studies of Science* 14 (1984): 97–113.

Jopling, D. G., Gage, S. J., and M. E. F. Schoeman. "Forecasting Public Resistance to Technology: The Example of Nuclear Power Reactor Siting." In J. R. Bright and M. E. F. Schoeman (eds.), *A Guide to Practical Technological Forecasting.* Englewood Cliffs, N.J.: Prentice-Hall, 1973, pp. 53–66.

Kahoe, R. D. "Motivations for Urban-Rural Migration." *Journal of Social Psychology* 96, 2 (1975): 303–4.

Kenard, B. *Nothing Can Be Done, Everything Is Possible.* Andover, Mass.: Brick House Publishing Co., 1982.

King, L. R. and P. H. Melanson. "Knowledge and Politics: Some Experiences from the 1960's." *Public Policy* 20 (Winter 1972): 82–101.

Klessig, L. L. and V. L. Strite. *The ELF Odyssey: National Security Versus Environmental Protection.* Boulder, Colo.: Westview Press, 1980.

Krannich, R. S. "Socioeconomic Impacts of Power Plant Development on Nonmetropolitan Communities: An Analysis of Perceptions and Hypothesized Impact Determinants in the Eastern United States." *Rural Sociology* 46 (Spring 1980): 128–42.

Lokken, L. F. *Michigan's Forest Products Industry Development Program: A Progress Report.* Lansing, Mich.: Office of the Governor, May 1985.

Long, L. H. and K. A. Hansen. *Reasons for Interstate Migration: Jobs, Retirement, Climate, and Other Influences.* Current Population Reports, Special Studies Series P–23, no. 81. U.S. Bureau of Census, Washington, D.C., March 1979.

Lovins, A. B. "Energy Strategy: The Road Not Taken." *Foreign Affairs* 55, 1 (1976): 65–96.

———. "Soft Energy Technologies." *Annual Review of Energy* 3 (1978): 477–517.

Marans, R. W. and J. D. Wellman. *The Quality of Nonmetropolitan Living: Evaluations, Behaviors, and Expectations of Northern Michigan Residents.* Ann Arbor, Mich.: Survey Research Center, 1978.

Matheny, A. R. and B. A. Williams. "Knowledges vs. NIMBY: Assessing Florida's Strategy for Siting Hazardous Waste Disposal Facilities." *Policy Studies Journal* 14, 1 (1985): 70–80.

Mazur, A. *The Dynamics of Technical Controversy.* Washington, D.C.: Communications Press, 1981.

———. "Opposition to Technological Innovation." *Minerva* 13 (Spring 1975): 58–81.

McCarthy, K. F. and P. A. Morrison. "The Changing Demographic and Economic Structure of Nonmetropolitan Areas of the United States." *International Regional Review* 2, 2 (1977): 123–42.

Messing, M., O'Meara, M., and R. M. Hall. *Report on the Jurisdictional Authorities of State and Local Government Related to Centralized and Decentralized Alternative Energy Systems.* Environmental Policy Institute, Washington, D.C., November 1976.

Michigan, State of. *Michigan Wood Energy Development Plan: An Addendum to Michigan's Forest Resources, A Statewide Resources Plan.* Department of ·Commerce and Department of Natural Resources, Lansing, March 1986.

Milch, J. "The Toronto Airport Controversy." In D. Nelkin (ed.), *Controversy: The Politics of Technical Decisions.* 2nd ed. Beverly Hills, Calif.: Sage, 1984, pp. 27–50.

Milliken, W. G. "Keynote Address." In M. L. Hiser (ed.), *Wood Energy: Proceedings of Governor William G. Milliken's Conference, November 29, 1977, Ann Arbor, Michigan.* Ann Arbor, Mich.: Ann Arbor Science Publishers, 1978, pp. 11–12.

Mitchell, R. C. "Since Silent Spring: Science, Technology and the Environmental Movement in the United States." In H. Skoie (ed.), *Scientific Expertise and the Public: Conference Proceedings.* Oslo: Insitute for Studies in Research and Higher Education, Norwegian Research Council for Science and the Humanities, 1979, pp. 171–207.

Molotch, H. "Oil in Santa Barbara and Power in America." *Sociological Inquiry* 40 (Winter 1970): 131–44.

Moore, M. D. (ed.). *Proceedings of Governor James J. Blanchard's Conference on Forest Resources: Creating 50,000 New Jobs in Michigan Forest Products Industries, March 22–23, 1983, Michigan State University, East Lansing, MI.* Lansing, Mich.: State of Michigan, 1983.

Morrison, P. A. and J. P. Wheeler. "Rural Renaissance in America? The Revival of Population Growth in Remote Areas." *Population Bulletin* 31, 3 (1976), entire issue.

Motz, A. B. "Toward Assessing Social Impacts: The Diachronic Analysis of Newspaper Contents." In K. Finsterbusch and C. P. Wolf (eds.), *Methodology of Social Impact Assessment.* Stroudsburg, Pa.: Dowden, Hutchinson, and Ross, 1977, pp. 303–13.

Mulkay, M. *Science and the Sociology of Knowledge.* Boston: G. Allen and Uwin, 1979.

National Academy of Sciences. *Energy Choices for a Democratic Society.*

Supporting Paper 7. Consumption, Location, and Occupational Patterns Resources Group Synthesis Panel, Committee on Nuclear and Alternative Energy Systems (CONAES), Washington, D.C., 1980.

Nelkin, D. "Controversy as a Political Challenge." In B. Barnes and D. Edge (eds.), *Science in Context: Readings in the Sociology of Science*. Cambridge, Mass.: MIT Press, 1982, pp. 276–81.

———. (ed.). *Controversy: The Politics of Technical Decisions*. 2nd ed. Beverly Hills, Calif.: Sage, 1984.

———. "Nuclear Power and Its Critics: A Siting Dispute." In D. Nelkin (ed.), *Controversy: The Politics of Technical Decisions*. 2nd ed. Beverly Hills, Calif.: Sage, 1984, pp. 51–71.

———. *Nuclear Power and Its Critics: The Cayuga Lake Controversy*. Science, Technology and Society Series no. 1. Ithaca, NY: Cornell University Press, 1971.

———. "The Political Impact of Technical Expertise." *Social Studies of Science* 5 (1975): 35–54.

———. "The Role of Experts in a Nuclear Siting Controversy." *Bulletin of the Atomic Scientists* 30 (November 1974): 29–36.

———. "Science and the Polity: Changing Relationships and Their Consequences for Social Studies of Science: Presidential Address—Fourth Annual Meeting." *Society for Social Studies of Science Newsletter* 4, 4 (Fall 1979): 5–8.

———. "Science, Technology, and Political Conflict: Analyzing the Issues." In D. Nelkin (ed.), *Controversy: The Politics of Technical Decisions*. 2nd ed. Beverly Hills, Calif.: Sage, 1984, pp. 9–22.

———. "Scientific Knowledge, Public Policy, and Democracy: A Review Essay." *Knowledge: Creation, Diffusion, Utilization* 1, 1 (September 1979): 106–22.

———. "Thoughts on the Proposed Science Court." *Newsletter on Science, Technology, and Human Values* no. 18 (January 1977): 20–31.

———. and M. Pollak. "Ideology as Strategy: The Discourse of the Anti-Nuclear Movement in France and Germany." *Science, Technology and Human Values* 5 (Winter 1980): 3–13.

———. "Problems and Procedures in the Regulation of Technological Risk." In R. C. Schwing and W. A. Albers (eds.), *Societal Risk Assessment: How Safe Is Safe Enough*? New York: Plenum Press, 1980, pp. 233–48; discussion 248–53.

Nichols, K. G. "The De-Institutionalization of Technical Expertise." In H. Skoie (ed.), *Scientific Expertise and the Public: Conference Proceedings*. Oslo: Institute for Studies in Research and Higher Education, Norwegian Research Council for Science and the Humanities, 1979, pp. 35–48.

Nowotny, H. "Experts and Their Expertise: On the Changing Relationship between Experts and Their Public." *Bulletin of Science, Technology, and Society* 1 (1981): 235–41.

———. "The Role of the Experts in Developing Public Policy: The Austrian Debate on Nuclear Power." *Science, Technology, and Human Values* 5 (Summer 1980): 10–18.

Ohi, J. M. et al. *Decentralized Energy Studies: Bibliography*. SERI/RR–774–448. Solar Energy Research Institute, Golden, Colo., May 1980.

Pimental, D., Chick, S., and W. Vergara. "Energy from Forests: Environmental and Wildlife Implications." *Interciencia* 6 (1981): 329–35.

Ploch, L. A. "The Reversal in Migration Patterns—Some Rural Development Consequences." *Rural Sociology* 43, 2 (Summer 1978): 293–303.

President's Commission for a National Agenda for the Eighties. *Science and Technology: Promises and Dangers in the Eighties.* Washington, D.C.: U.S. Government Printing Office, 1981.

Primack, J. and F. von Hippel. *Advice and Dissent: Scientists in the Political Arena.* New York: Basic Books, 1974.

Rathge, R. W. *The Institutional Impacts of Rapid Population Growth on a Nonmetropolitan Michigan County.* Ph. D. dissertation. Michigan State University, East Lansing, Mich., 1981.

Ravetz, J. R. "Scientific Knowledge and Expert Advice in Debates about Large Technological Innovations." *Minerva* 16, 2 (Summer 1978): 273–82.

Rayner, J. and D. Peerla. "To Spray or Not to Spray? Exclusive Concepts of Science and Nature in the Canadian Spruce Budworm Controversy." Paper presented at the First National Symposium on Social Science in Resource Management, Corvallis, Oreg., May 1986.

Reich, M. R. "Environmental Politics and Science: The Case of PBB Contamination in Michigan." *American Journal of Public Health* 73, 3 (March 1983): 302–13.

Richter, K. *Nonmetropolitan Growth in the Late 1970s: The End of the Turnaround?* Working Paper 83–20. Center for Demography and Ecology, University of Wisconsin, Madison, Wis., 1983.

Rip, A. "Controversies as Informal Technology Assessment." *Knowledge: Creation, Diffusion, Utilization* 8, 2 (1986): 349–71.

Rolka, H. and P. E. Sworden. "SECO—Dow Corning's Wood Fired Industrial Cogeneration Project." In *Advances in Energy Productivity: Proceedings of the 5th World Energy Congress, September 14–17, 1982.* Association of Energy Engineers, 1982, pp. 147–50.

Rozak, T. "Citadel of Expertise." Chap. 3 in *Where the Wasteland Ends.* Garden City, N.Y.: Anchor Books, 1973, pp. 28–73.

Sclove, R. "Decision-making in a Democracy," *Bulletin of the Atomic Scientists* (May 1982): 44–49.

Sells, W. H. "Use of Wood as an Energy Source." In *Proceedings of the 1979 Energy Information Forum and Workshop for Educators.* Lansing, Mich.: Michigan Educators Energy Forum, 1979, pp. 91–95.

Shama, A. and K. Jacobs. *Social Values and Solar Energy Policy: The Policymaker and the Advocate.* Solar Energy Research Institute, Golden, Colo., 1980(?).

Sokolow, A. "Local Politics and the Turnaround Migration: Newcomer-Oldtimer Relations in Small Communities." In C. Roseman, A. Sofranko, and J. Williams (eds.), *Population Redistribution in the Midwest.* North Central Regional Center for Rural Development, Iowa State University, Ames, 1981, pp. 169–90.

Stokes, B. *Helping Ourselves: Local Solutions to Global Problems.* Washington, D.C: Worldwatch Institute, 1981.

Tichenor, P. J., Donohue, G. A., and C. N. Olien. *Community Conflict and the Press.* Beverly Hills, Calif.: Sage Publications, 1980.

Tierney, S. F. "The Nuclear Waste Disposal Controversy." In D. Nelkin (ed.), *Controversy: The Politics of Technical Decisions.* 2nd ed. Beverly Hills, Calif.: Sage, 1984, pp. 91–110.

Tribe, L. H., Schelling, C. S., and J. Voss (eds.). *When Values Conflict: Essays on*

Environmental Analysis, Discourse, and Decision. Cambridge. Mass.: Ballinger Publishing Co., 1976.

U.S. Congress, Senate. Select Committee on Small Business and the Committee on Interior and Insular Affairs. *Alternative Long Range Energy Strategies: Appendices.* 94th Congress, 2nd Session. Washington, D.C.: U.S. Government Printing Office, 1977.

——. *Alternative Long Range Energy Strategies: Hearings, December 9, 1976, 94th Congress, 2nd Session.* Washington, D.C.: U.S. Government Printing Office, 1977.

U.S. Department of Energy. *Environmental Readiness Document—Wood Combustion.* DOE/ERD-0026. Assistant Secretary for the Environment, Washington, D.C., August 1979.

U.S. Environmental Protection Agency. *Preliminary Environmental Assessment of Biomass Conversion to Synthetic Fuels.* EPA-600/778-204. Battelle Columbus Laboratory for EPA/IERI, Cincinnati, Ohio., October 1978.

Voss, P. R. and G. V. Fuguitt. *Profile: The Region's New Residents.* Upper Great Lakes Regional Commission, U.S. Department of Commerce, Washington, D.C., December 1979.

Veigel, J. and J. H. Lohnes. "The Forest and Energy." In C. E. Hewett and T. E. Hamilton (eds.), *Forests in Demand: Conflicts and Solutions.* Boston, Mass.: Auburn House Pub., 1982, pp. 43–50.

Vietor, R. H. K. *Environmental Politics and the Coal Coalition.* College Station, Tex.: Texas A & M University Press, 1980.

Whittemore, A. S. "Facts and Values in Risk Analysis for Environmental Toxicants." *Risk Analysis* 3, 1 (March 1983): 23–33.

Wilkes, J. M. "Case Studies: A Promising Way to Assess Technological Impacts?" *4S Review* 1, 2 (Summer 1983): 8–21.

Williams, J. D. and A. J. Sofranko. "Motivations for the Immigration Component of Population Turnaround in Nonmetropolitan Areas." *Demography* 16, 2 (1979): 239–55.

Williams, R. M., Jr. *American Society.* 3rd ed.. New York: Knopf, 1970.

Wolf, C. P. "The NIMBY Syndrome: Its Cause and Cure." In F. Sterrett (ed.), *Environmental Sciences.* New York: New York Academy of Sciences, 1987, pp. 216–29.

Yin, R. K. *Case Study Research: Design and Methods.* Volume 5. Applied Social Research Methods Series. Beverly Hills, Calif.: Sage, 1984.

Newspaper, Newsletter, and Magazine Articles

Applegate, J. "Ultrasystems Pins Hopes on Wood Power." *Los Angeles Times* sec. IV, 1, 4.

Anderson, K. "Burlington's Wood and Water." *Public Power* 39, 6 (1981): 41–44.

Big Rapids Pioneer. "Farwell to Host Wood Meeting." 17 June 1980, 1A.

Biologue: Biomass Energy News and Information. "Burlington Electric Department Problems Continue." 3, 2 (1986): 2, 18.

Biologue: Biomass Energy News and Information. "Wood Fired Plants Facing Opposition." 3, 2 (1986): 1.

Black, K. "Tilting at Windmills." *Northwest Magazine* 18, 1 (1987): 13–16.

Blackman, T. "Another Wood-fueled Plant Fires Up Its Generators." *Forest Industries* 113, 10 (1986): 23.

———. "Mechanized Thinning Helps Forest, Power Production." *Forest Industries* 112, 10 (1986): 24–25.

Bolyard, D. "Letters to the Editor: Hersey Plant." *Michigan Out-of-Doors* 35 (May 1979): 6, 8.

Deis, R. "Where There's Wood, There's Smoke: Are We Being Burned by Burning Wood?" *Environmental Action* 12 (December 1980): 4–9.

Detroit Free Press. "Industrial-Strength Wood Powers Is Back." 2 March 1984, 1C.

———. "Town Debates Wood-Power Plant." 13 February 1979, 3, 11A.

Detroit News. "Cash in the Chips: Firm Sees Big Savings on Energy in New Wood-Fueled Power Plant." 18 January 1983, 4B,

———. "Electric Plant Bid Sets Off Sparks." 15 March 1984, 1, 2F.

———. "Power Play: Firms Drop Utilities, Create Their Own Energy." 24 January 1985, 1–2A.

———. "Tiny Town Beats Power Plant Bid by Two Utilities." 18 May 1980, 5B.

———. "Wood as Energy." 28 January 1983, 12A.

Evart Review. "Reagan Lauds Morbark." 4 January 1979, 14.

Frank, E. P. "The Yankee Forest: Will It Be Plundered or Preserved by Wood Heat?" *New Roots*, no. 9 (January–February 1980): 29–32.

Gove, B. "The Saga of Burlington Electric: The Current Status of What Was To Be a Model Wood Energy System." *Northern Logger and Timber Processor* 35 (November 1986): 26–30.

———. "Wood-fueled Power Plants Promise Uncertain Future for Vermont / New Hampshire." *Northern Logger and Timber Processor* 36 (November 1987): 22–23.

Graff, G. "Wood-fueled Cogeneration: Keeping a Good Thing Going in Michigan." *Northern Logger and Timber Processor* 37 (November 1988): 26–29.

GRP. "Hersey Does Slow Burn over Proposed Wood-Burning Plant." 21 January 1979, 1E.

———. "Judge Halts Plans for Cheboygan Wood-Burning Plant." 23 November 1984, 13A.

———. "Many Towns Want Wood Electric Plant." 25 May 1980, 13A.

———. "Ogemaw Beckons Waste-Fueled Power Plant." 27 June 1980, 14A.

———. "Osceola Will Get Wood-Burning Power Plant." 13 January 1979, 8B.

———. "Wood-Chip Fuel Gets a Warm Reception at Dow Corning's New Plant in Midland." 18 January 1983, 7A.

———. "Wood-Fueled Power Plant Idea Is Debated." 3 August 1980, 13A.

Abbreviations used are *GRP* for the *Grand Rapids Press; LSJ* for the *Lansing State Journal;* and *OCH* for the *Osceola County Herald*.

———. "Wood-Fueled Power Plant Is Put on Shelf." 11 September 1980, 12A.

Hacker, D. "Three Villages Branch Out by Planning Wood Power." *Detroit Free Press*, 13 October 1987, 3, 10A.

Hale, J. "Electric Generating Plant under Study." *Big Rapids Pioneer*, 6 June 1978, 1A.

Harris, M. "The Boom in Wood Use: Promise or Peril? *American Forests* 86 (1980): 34–36+.

———. "Vermont: Goodbye Coal, Hello Trees." *Mother Jones* 3 (December 1978): 13–14.

Institute for Social Research Newsletter. "Local Officials Must Plan for Growth, Northern Michigan Residents Declare." 6, 3 (Summer 1978): 2, 8.

Johnson, E. "Biomass Energy: Caught in the Middle of New Hampshire." *Northern Logger and Timber Processor* 37 (November 1988): 6–7.

LSJ. "Electricity from Wood Must Wait." 12 September 1980, 8B.

———. "Loggers, Environmentalists Fight Forest Plan." 12 December 1986, 3B.

———. "New Wood Processes Step Up Timber Needs, Spark Battle over Land." 19 October 1986, 1E.

———. "1983 Target for Wood-Fired Plant." 6 February 1979, 4B.

———. "North Michigan Town Finds Key to Revitalization." 16 November 1986, 7B.

———. "Plant Powered by Wood." 18 January 1983, 2B.

———. "Residents Split on Energy Plant." 18 March 1984, 7B.

Makansi, J. "Wood-Burning Cogen System Makes Debut in the Chemical Industry." *Power* 128 (February 1984): 103–4.

Miller, B. "Resourceful People." *Michigan Natural Resources* 48, 5 (September–October 1979): 42–47.

Michigan Out-of-Doors. "Wood To Provide Energy for Small Town." (May 1978): 113.

Mining Journal. "Consumers' Woes Hurting Small Power Plants." 9 April 1985, 5A.

Mitchell, R. "From Elite Quarrel to Mass Movement." *Society* 18, 5 (1981): 76–84.

New York Times. "Power Plant in Vermont Fueled by Wood Chips." 16 April 1984, 14A.

Northern Logger and Timber Processor. "Despite Oil Price Plunge, Wood Energy Market Looks Strong—For the Moment at Least." 34 (March 1986): 1.

———. "Environmentalist Groups Sue over Proposed Maine Biomass Plant." 36 (November 1987): 2.

———. "More than 20 Biomass Energy Projects Planned for Maine." 33 (January 1985): 8.

———. "New England Wood Energy Situation Brightens Up." 35 (January 1987): 1.

———. "Town Planning Commission Approves Biomass-Burning Plant in Ryegate, VT." 35 (February 1987): 2.

———. "Ultrapower Hits Yet Another Potential Snag in New York State." 34 (October 1985): 2.

———. "Upstate New York Plant Proposals under Scrutiny." 33 (January 1985): 8.

———. "Vermont's Burlington Electric under Fire." 34 (January 1986): 4.

———. "Wood-fired Power Plants in Maine's Aroostook County To Be Built." 36 (June 1988): 2.

OCH. "Board Commended." 27 December 1979, 4A.

———. "Commissioners Urged to Reconsider." 8 November 1979, 4A.

———. "CPC to Delay Wood Plant Study: Mitchell Concerned over Decision." 18 September 1980. 1A.

———. "CRUF Is Not Short on Facts." 5 June 1980, 2A.

———. "CRUF Reaffirms Stand Opposing Hersey Wood Chip Plant." 25 October 1979, 1A.

———. "The Demise of the Hersey Woodburner—Why and How?" 31 July 1980, 5B.

———. "DNR Delays Wood Decison." 13 September 1979, 3A.

———. "Fight Facts Were Presented." 8 May 1980, 2A.

———. "Hersey Residents Not Sold on Plant." 21 December 1978, 1A.

———. "Hersey Residents Plan Meeting." 14 December 1978, 1A.

———. "Hersey Site Chosen for Chipping Plant." 28 September 1978, 1A.

———. "Hersey Woodchip Plant Monster." 15 February 1979, 4A.

———. "Local Towns Offer Sites for Plant." 19 June 1980, 1A.

———. "Opinions Split on Plant Near Hersey." 26 October 1978, 1, 3A.

———. "RDF Hot Issue in Hersey." 17 April 1980, 1, 4B.

———. "Scrambling to the Need." 29 November 1979, 4A.

———. "200 Attend Hersey Wood Chip Plant Meeting." 15 February 1979, 1A.

———. "Vermont Wood Chip Plant Shelved." 15 November 1979, 1A.

———. "What's Really Being Planned for the Hersey Area?" 18 October 1979, 4A.

———. "The Wood Chip Controversy: Pro: Money's A Key ... Con: Use Is Inefficient." 12 April 1979, 3, 8B.

———. "Wood Chip Executive Speaks Out: Hersey Still Probable Site." 31 January 1979, 3, 5A.

———. "Wood Chip Plant Fuel Supply Approved for 10 Years." 18 October 1979, 1A.

———. "Young Explains Success of Vermont Plant." 25 January 1979, 7A.

———. "Your Choice: Fact or Fiction." 24 April 1980, 4A.

Plant Engineering. "Woodburning Cogeneration System Cuts Plant Energy Costs." 38 (9 February 1984): 26+.

Pollution Engineering. "Woodburning Cogeneration Power Plant Completed on Time, Near Budget." 16 (April 1984): 6.

Ponczak, G. "University Converts Heat Plant in Order To Use Local Timber." *Energy User News*, 11 February 1985, 2–3.

Reason, J. "Wood-fired Powerplants Are Not Without Problems." *Power* 130 (August 1986): 66–67.

Rinebolt, D. C. "Wood Power for the Future." Public Power 46, 6 (1988): 37–39.

Index